NDE-Vol. 13

NDE FOR THE ENERGY INDUSTRY 1995

presented at
THE ENERGY AND ENVIRONMENTAL EXPO '95 —
THE ENERGY–SOURCES TECHNOLOGY CONFERENCE AND EXHIBITION
HOUSTON, TEXAS
JANUARY 29–FEBRUARY 1, 1995

sponsored by
THE NONDESTRUCTIVE EVALUATION DIVISION, ASME

edited by
DON E. BRAY
TEXAS A&M UNIVERSITY

THE AMERICAN SOCIETY OF MECHANICAL ENGINEERS
345 East 47th Street ■ United Engineering Center ■ New York, N.Y. 10017

ISBN No. 0-7918-1298-7

Library of Congress Catalog Number 94-74526

FOREWORD

Safe and cost effective operation of our nation's infrastructure depends on nondestructive evaluation (NDE). Programs for life extension and fitness-for-service analyses for operating systems, and for assuring quality components and systems in the manufacturing arena, require well-based estimates of defect size and orientation.

Moreover, knowledge of material characteristics such as residual stress, hardness, and heat treatment can no longer be neglected simply because they are thought to be unmeasurable. Energy companies and their suppliers are heavy users of nondestructive evaluation. Also, energy consumers in the various areas of transportation can benefit from the efficient and weight saving designs that occur if NDE is a key participant in the initial design and material selection process. The papers presented in this NDE Symposium will include topics ranging from uncertainty in flaw size estimation to nondestructive stress measurement. Session one contains papers on erratic measurement, personnel qualification, and transducer calibration. Session two papers describe inspections of vessels, tanks, and structures. Several applications on residual stress measurement are described in session three. Finally, in session four typical applications for structures and components are discussed. We are pleased to present these papers describing NDE contributions to improved efficiency and safe operation throughout the energy industry. It is our hope that these presentations will provide a forum for discussion of NDE topics that will lead to advancements in applications to our industry.

I am happy to acknowledge the signficant contributions of the session chairmen and co-chairmen who put forth extraordinary efforts to see that this proceedings contains well organized and reviewed papers from their sessions. The reviewers also made diligent and useful comments on the papers which, in each case, improved the quality of the manuscript. Further, I appreciate the authors who have prepared papers that are both timely and informative.

Don E. Bray
Texas A&M University

SESSION CHAIRMEN AND CO–CHAIRMEN

Glenn Armstrong, Engineering Design and Testing Corporation, Houston, Texas
Russel Austin, Texas Research Institute, Austin, Texas
Rathi Bhatacharya, Bradley University, Peoria, Illinoia
Mike Brauss, Proto Manufacturing, Limited Oldcastle, Ontario, Canada
Glenn Cunningham, Center for Electric Power, Tennessee Tech University,
    Cookeville, Tennessee
Bill Friedman, Knoll Atomic Power Laboratory, Schenectady, New York
Paul G. Junghans, Atlas Wireline Services, Houston, Texas
Rod K. Stanley, NDE Information Consultants, Houston, Texas

LIST OF REVIEWERS

G. Armstrong	B. Friedman
R. Austin	P. Junghans
R. Bhatacharya	A. Holt
J. Bower	K. Kozaczek
A. Bray	M. Longnecker
D. Bray	R. Pfluger
J. Cook	A. Porter
G. Cunningham	J. Porter
C. Darvennes	M. Srinivasan
J. R. Dickens	R. Stanley
S. Elhoual	R. Streit

CONTENTS

ERRATIC MEASURE

Terry Oldberg
Consultant
Los Altos Hills, California

Ronald Christensen
Entropy Limited
Lincoln, Massachusetts

ABSTRACT

The paper exposes an inconsistency in the field of non-destructive examination (NDE) and points out the dangers of the confusion which is created by it. The physical objects which occupy a statistical population occupy a partition of the complete set of physical objects under consideration. However, many of NDE's procedures test physical objects which do not occupy such a partition. The inconsistency is set up when NDE's scientists represent physical objects which do not occupy partitions as the elements of populations in their studies of NDE's reliability. The bogus populations invalidate the definition of probability as a measure of an event whose value on a certain event is 1.

With probability invalidated as a measure of a test's reliability, NDE's scientists have proceeded by representing a different measure of an event as a "probability." The value of this measure on a certain event varies between 0 and 2 in one U. S. Nuclear Regulatory Commission study. This makes the reader who interprets the study's measure as a probability wrong on a number of counts. For example, the apparently perfect certainty which accompanies a value of 1 for the study's Probability of Detection is actually perfect ambiguity because this measure's value on the certain event of *a defect* is 2. Dire consequences are imaginable if a reactor engineer were to act on the USNRC's representation.

The authors recommend avoidance of the consequences of this and other instances of confusion by the translation of past reporting. The translated reports would use a language which discriminated proper from improper probabilities and populations. For the longer term, they recommend the refocusing of NDE to yield only proper probabilities and populations.

INTRODUCTION

Probability is a measure of an event whose value on a certain event is 1 (Halmos, 1950). However, certain irregularities in the design of a study's statistical population can yield a "probability" whose value on a certain event varies significantly from 1. Like the yardstick whose "yard" varies significantly from the orthodox 0.9144 meters, this "probability" has the capacity to seriously mislead people. That it has been the unsuspected measure of the reliability of non-destructive examination (NDE) in error-sensitive areas of engineering prompts the following warning.

Engineers use the radiographic testing, ultrasonic testing and other methods of NDE to diagnose problems with structures which function under mechanical stress. They are particularly apt to use it in situations in which the failure of a structure might cause harm. However, NDE can itself cause harm when it is in error. Thus, engineers have been moved to establish the probabilities of error in the various methods of NDE. These "probabilities" are sometimes not probabilities.

This paper has three parts. The first part, **Probability versus Pseudoprobability**, establishes and contrasts two measures of an event. Probability is consistent with population while pseudoprobability is consistent with irregular or *pseudo-population*. While both measures are legitimate, they are different measures. Therefore, when the literature labels a study's pseudoprobability as a "probability," its reader can be seriously misled. The second part, **Nuclear Confusion**, exposes this kind of confusion in a U. S. Nuclear Regulatory Commission study. In relation to dangerous levels of damage to a nuclear reactor component, its Probability of Detection is a pseudoprobability whose value on the certain event of *a defect* is 2. Thus, the value of 1 which the agency reports for the "probability" of detecting a dangerous condition in a nuclear reactor suggests perfect certainty that the test is valid but is actually perfect ambiguity regarding whether the test is valid. The third part, **Conclusions and Recommendations**, warns readers to expect similar surprises throughout NDE's literature. It recommends that NDE's literature be translated into the language of pseudoprobabilistics so that it does not mislead people. The paper closes by urging that NDE be restructured to define actual populations and probabilities.

PROBABILITY VERSUS PSEUDOPROBABILITY

In *Sampling Techniques*, the statistician William Cochran provides a rule which yields a population when it is followed. Before a study's sample may be drawn, its population must be divided into the physical objects which are called *units*. The units must "... cover the whole of the population and they must not overlap, in the sense that every element in the population belongs to one and only one unit (Cochran, 1977)." Cochran is stating that a study's units must occupy a partition of their population. In the remainder of the paper, we will endeavor to explain what this means.

A class $\{A_1, ..., A_n\}$ of sets $A_1, ..., A_n$ is said to be a *partition* of set A if every element of A belongs to a set in $\{A_1, ..., A_n\}$ and no two sets in $\{A_1, ..., A_n\}$ have an element in common. Figure 1 shows a class of two triangles which is a partition of a rectangle. In this example of a Venn diagram, a class is represented by a set of geometrical objects, a set by a single geometrical object and an element of a set by a point within a geometrical object's boundary.

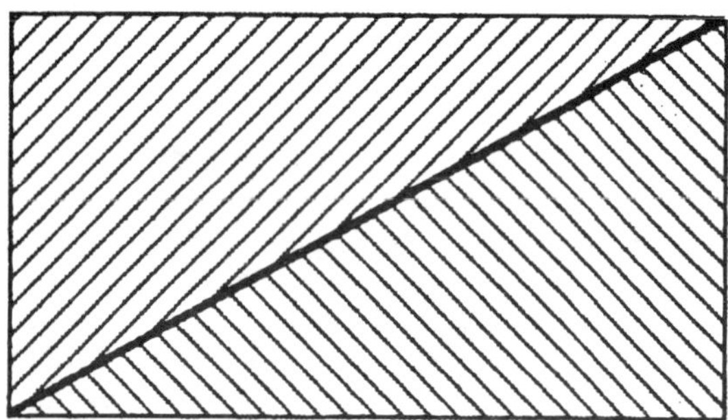

Figure 1 A partition. The class of two triangles is a partition of the rectangle.

We will reference a class $\{A_1, ..., A_n\}$ as a *pseudopartition* of a set A if it is not a partition of A. Figure 2 shows an example of a pseudopartition. The class of two circles is a pseudopartition of the rectangle. Note the region of overlap between the two circles and the region of the rectangle that is not covered by any circle.

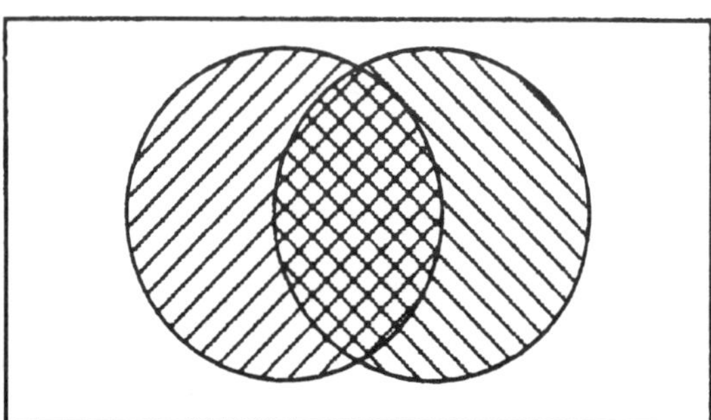

Figure 2 A pseudopartition. The class of two circles is a pseudopartition of the rectangle.

In this paper, the entity which Cochran terms an "element in the population" will be termed a *data point*. In the study of testing reliability, each data point is a pair of numbers representing the tested and true values of the property of a physical object which is measured by the test. The complete set of possible data points is termed a study's *sample space*. An *event* is a subset of a sample space.

In characterizing the reliability of a test, the test's investigator selects a partition of the sample space. Then he estimates a value for the probability of each event in the partition. Though a variety of partitions are available, investigators usually select the simplest. Its events are termed *a true positive, a false negative, a true negative* and *a false positive*. Figure 3 shows the sample space which pertains to the study of testing reliability and this partition of it.

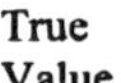

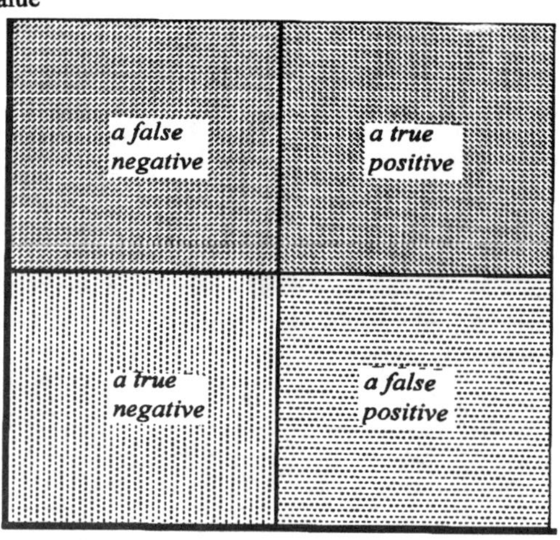

Figure 3 The sample space which pertains to the study of testing reliability and the most popular partition of it.

The *relative frequency* of an event is the number of physical objects which participate in the event divided by the number of data points in the sample space. The probability of an event models the relative frequency of the same event. Conversely, the value of the relative frequency of an event provides empirical verifiability for the value which some theory claims for the probability of this event. Empirical verifiability is the hallmark of science (Popper, 1959).

From the standpoint of empirical verifiability, a certain event has particular significance, for the value of its probability is 1 by the definition of *probability*. The relative frequency of a certain event is the number of physical objects which participate in the event divided by the number of data points in the event. For example, in the testing of patients for the virus Hepatitis B, the relative frequency of the certain event of *a true positive* is the number of true positive patients divided by the number of true positive outcomes. A relative frequency of a certain event whose value is 1 is consistent with the definition of *probability*. A relative frequency of a certain event whose value is not 1 is inconsistent with the definition of *probability*.

The relative frequency of a certain event is a measure of the degree of coverage of this event by physical objects. When the relative frequency of every event in a partition of a study's sample space is 1, this signifies that each physical object in the complete set of them participates in one and only one event. If this is true, the class of sets of physical objects which participate in the various events is a partition of the complete set of physical objects.

There are two other possibilities. In the first possibility, there are physical objects which participate in more than one event in a partition of the sample space. One says, in this case, that the sets of physical objects corresponding to these events *overlap*. In the second possibility, fewer physical objects participate in an event than there are data points. One says, in this case, that the set of physical objects corresponding to this event *under covers* the event. If the class of sets of physical objects corresponding to the partition of a sample space exhibits under coverage or overlap, it is a pseudopartition of the complete set of physical objects.

A pseudopartition of the complete set of physical objects is inconsistent with the empirical verifiability of a probabilistic model whereas a partition is consistent with the empirical verifiability. As empirical verifiability is the hallmark of science, great scientific importance attaches to differentiating the pseudopartition from the partition. In normal statistics, one terms a physical object which belongs to a partition a *unit*, a set of them a *population* and a subset of a population a *subpopulation*. No uniform terminology has emerged in abnormal statistics so we will coin a terminology. We will term a physical object which belongs to a pseudopartition a *pseudounit*, a set of them a *pseudopopulation* and a subset of a pseudopopulation a *pseudosubpopulation*. To complete the analogy, we define *pseudoprobability* as a measure of an event whose value on a certain event equals the value of the relative frequency of this event. The value of the pseudoprobability of a certain event is empirically verified by the relative frequency of the same event. Thus, a pseudopartition of a complete set of physical objects is consistent with the empirical verifiability of a pseudoprobabilistic model.

The physicist Lazar Mayants has noted that the discipline which includes the concepts of probability, partition, population, subpopulation and unit is empirically verifiable and can therefore be considered a science (Mayants, 1984). He terms this science *probabilistics*. It may be noted that the discipline which includes the concepts of pseudoprobability, pseudopartition, pseudo-population, pseudosubpopulation and pseudounit is also empirically verifiable and thus also a science. We will term this science *pseudoprobabilistics*.

Though probabilistics and pseudoprobabilistics are each internally consistent, their concepts cannot be mixed. Thus, it is important to keep the concepts separate linguistically.

One way in which the concepts can be confused is by the representation of pseudoprobability as "probability." This kind of confusion has arisen in a study of the U. S. Nuclear Regulatory Commission (USNRC).

NUCLEAR CONFUSION

E. R. Bradley and his colleagues studied the reliability of NDE in a nuclear power reactor's steam generator tubes (Bradley et al., 1988). The tubes have the import of containing a reactor's cooling water while transferring the reactor's heat to the outside. Possible consequences from the rupture of tubes include the meltdown of the affected reactor's core and release of its huge inventory of radioactive materials into the biosphere.

Nonetheless, the tubes have proven susceptible to weakening by corrosion. A number of tubes have burst during reactor operations and the USNRC has reacted by imposing periodic inspections of the tubes by NDE. Bradley's has been the only substantial study of the reliability of these inspections.

Each tube's inspector scans it with a remote sensor. The sensor is inserted into the bore of each tube which has been slated for inspection. Then it is translated along this tube's axis. As it moves, the sensor emits a stream of data. The data stream is decoded by the inspector.

Decoding produces a set of discrete *indication*s of damage. Each indication contains an estimate of the radially directed penetration of the tube by corrosion, the identity of the tube and the axial position of the sensor. The indications are recorded.

Bradley asked various teams of inspectors to test a prescribed set of tubes under the rules by which they would be inspected in practice. These rules had been written by the American Society of Mechanical Engineers (ASME) and enacted by the USNRC. Then he and his colleagues dismantled the tubes in an attempt at establishing the reliability of their inspection. In modeling the reliability, Bradley selected the usual sample space partition, that is, the one containing *a true positive, a false negative, a true negative* and *a false positive*.

In the mechanical trades and in statistics, the event of *a true positive* or *a false negative* is often termed *a defect*. The event of *a true negative* or *a false positive* may be termed *a non-defect* by extension. The two events of *a defect* and *a non-defect* form a partition of the sample space in the study of testing reliability.

A simplification arises when *a defect* and *a non-defect* are taken to be certain. With *a defect* certain, the probability or pseudoprobability of *a true positive* and the probability or pseudoprobability of *a false negative* sum to the relative frequency of the certain event of *a defect*. As the two probabilities or pseudoprobabilities are dependent, the reliability of a test may be characterized in terms of only one of them. Bradley bases his characterization on the probability or pseudoprobability of *a true positive*. He terms this measure the Probability of Detection.

Similarly, with *a non-defect* certain, the probability or pseudoprobability of *a true negative* and the probability or pseudoprobability of *a false positive* sum to the relative frequency of the certain event of *a non-defect*. As the two probabilities or pseudoprobabilities are dependent, the reliability of a test may be characterized in terms of only one of them.

In reading Bradley's report, one notices an oddity in the use of terminology which clouds understanding of whether the complete set of physical objects under the ASME-USNRC test contains units or pseudounits. The physical objects are identified as "defects" but *a defect* is the name of an event and not a physical object, as several statisticians take pains to tell their readers (Bowker et al., 1972; Juran, 1974).

However, in a recent paper Perdijon observes that in NDE, "defect" often designates an object which has a definite boundary in space. He suggests that the term "discontinuity" be reserved for such an object (Perdijon, 1993). In certain cases, a discontinuity's boundary surrounds some material. In these cases, a physical object is defined by the boundary.

This conception of the "defect" fails, however, to resolve the question, for Bradley's "defects" are nearly all the voids which are formed on the exterior of a tube by corrosion. A void is a definite boundary surrounding no material and is not physical. Such an object possesses no properties of its own for NDE to measure.

If they have no properties, however, Bradley's "defects" do not support the establishment of the property values which Bradley reports. Thus, we gather that Bradley is imputing to the discontinuities the properties of the physical objects in which they are embedded. However, we conclude, there is so little difference between these physical objects and the discontinuities that Bradley does not feel obliged to observe a distinction. According to this theory, Bradley's "defect" is a discontinuity's boundary plus a thin layer of material which adheres to its exterior. The layer is thin enough that the dimensions of a "defect" are essentially the same as the dimensions of the discontinuity which is embedded in it. However, this "defect" is a physical object because of the material. We will adopt this theory of Bradley's "defect" in proceeding with our analysis of his findings, while placing quotation marks around the word to remind our readers that it designates a physical object and not an event. When the word designates an event, it will be represented in italics as *a defect*.

Bradley's "defects" were defined by metallurgists performing destructive tests after the tubes had been inspected. The 108 "defects" were quite short, with a median, axially directed length of 1 inch. At ± 3 inches, the axially directed positional uncertainty of each indication was quite long by comparison. Thus, every indication falling on a "defect" might not have referenced this "defect." Conversely, every indication falling off a "defect," but within 3 inches of it, might have referenced this "defect." Because of the great positional uncertainty, then, the question of whether an indication falling within 3 inches of the end of a "defect" referenced this "defect" or did not reference it had a completely ambiguous answer.

Bradley dealt with this problem in way which is highly significant to the theme of this paper. If a "defect" had an indication within 3 inches of it, he assigned it to *a true positive*. Otherwise, he assigned it to *a false negative*. He then divided the number of "defects" he had assigned to *a true positive* by the total number of "defects" and assigned this ratio to the Probability of Detection.

A result of this methodology is presented in Bradley's report as a graph with the Probability of Detection as the ordinate and Metallurgical Wall Loss as the absissa. Metallurgical Wall Loss is the degree of penetration of a tube's wall thickness by a "defect." This graph is presented as this paper's Figure 4.

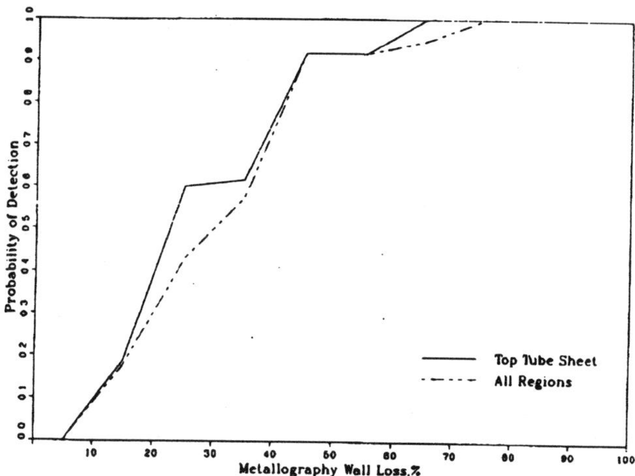

Figure 4   Graph of Probability of Detection versus Metallurgical Wall Loss (Bradley et al., 1988.)

Figure 4 shows the Probability of Detection rising from 0, when the Metallurgical Wall Loss is less than 5% to 1, when the Metallurgical Wall Loss is more than 75%. As tubes fail under service conditions only when the Metallurgical Wall Loss is more than 80%, this graph represents certainty that a "defect" will be indicated when it is at a dangerous level of corrosion. However, this representation conflicts with our earlier observation: whether an indication references a particular "defect" or something else is totally ambiguous. Let us analyze how this conflict arises within the framework of probabilistics and pseudoprobabilistics.

Bradley's probability estimate results from a methodology in which he assigns a "defect" to *a false negative* if he does not assign it to *a true positive*. This methodology is consistent with the assumption that the ASME-USNRC test defines a partition of the complete set of physical objects under testing. Bradley suggests this to be so, for he states that

An inspected unit of material can always be placed into one of the following categories:
- true positive (a defect indication was reported and there was an actual defect present)
- false positive (a defect indication was reported and there was no defect)
- false negative (no defect was reported and an actual defect was present)
- true negative (no defect was reported and no defect was present)

This is a description of a partition.

However, Bradley's rule of evidence produces a conflict with this assumption. It assigns every "defect" to *a true positive* if it is within 3 inches of an indication. The same rule assigns every such "defect" to *a false negative* also. Let us examine the consequences of this ambiguity.

Bradley's Probability of Detection (see Figure 4) is either the probability of *a true positive* or the pseudoprobability of *a true positive*. It is a probability if the relative frequency of the certain event of *a defect* is 1 and a pseudoprobability otherwise. Let n_{tp} designate the number of "defects" Bradley assigned to *a true positive* and n_{fn} the number he assigned to *a false negative*. Let f_d designate the relative frequency of the certain event of *a defect*. Then

$$f_d = \frac{2 \times n_{tp} + n_{fn}}{n_{tp} + n_{fn}}. \qquad (1)$$

We have doubled the number of "defects" Bradley assigned to *a true positive* in computing the number of physical objects corresponding to *a true positive* or *a false negative* because Bradley's rule of evidence assigns a "defect" to *a false negative* each time it assigns one to *a true positive* but Bradley did not make the assignment to *a false negative*.

Bradley estimates the Probability of Detection in Figure 4 from

$$\text{Probability of Detection} = \frac{n_{tp}}{n_{tp} + n_{fn}}. \qquad (2)$$

Combining Equations (1) and (2), one concludes that the relative frequency of the certain event of *a defect* is given by

$$f_d = 1 + \text{Probability of Detection}.$$

One sees that the value of the relative frequency of the certain event of *a defect* rises from 1 (when the Probability of Detection is at its minimum value of 0) to 2 (when the Probability of Detection is at its maximum value of 1.) When the Probability of Detection has a value of 0, it is a probability. When the Probability of Detection has a value which is not 0, it is a pseudoprobability.

The condition that the Probability of Detection has a value of 1 is particularly pertinent as this is its value, according to Figure 4, at dangerous levels of damage. Under this condition, the Probability of Detection shown in Figure 4 is a pseudoprobability with a value of 2 on the certain event of *a defect*. Thus, the value of 1 for the Probability of Detection at dangerous levels of damage should be viewed as 50% of the value of the pseudoprobability of the certain event of *a defect* and not the 100% of the value of the probability of the certain event of *a defect* which Figure 4's choice of language implies.

50% of a pseudoprobability with a value of 2 on the certain event of *a defect* tells a far different story than 100% of a probability. In particular, the value of 2 for the pseudoprobability of the certain event of *a defect* indicates complete overlap of the pseudosubpopulations corresponding to *a true positive* and *a false negative*. The value of 1 for the Probability of Detection at dangerous levels of damage indicates nothing more than that 50% of the pseudounits corresponding to *a defect* are assigned by the ASME-USNRC test to *a true positive*. The other 50% are assigned by this test to *a false negative*. There is complete ambiguity regarding whether this test is valid in diagnosing dangerous levels of damage, but there seems from the use of language to be complete certainty that the test is valid. A reactor engineer who took the complete ambiguity of Figure 4 to be the complete certainty its language implies might well commit a disastrous error.

The remaining portion of Bradley's methodology relates to the certain event of *a non-defect*, that is, *a true negative* or *a false positive*. Bradley finds that there are numerous indications without "defects" within 3 inches and assigns them to *a false positive*. He assigns no object at all to *a true negative*. He suggests no value for any probability of *a false positive* or *a true negative* but suggests this to be inconsequential because "...the safety issue involved in NDE reliability does not depend on the amount of nondefective material correctly passed, but whether the defective material is identified."

Let us analyze these results from the standpoint of probabilistics and pseudoprobabilistics. Though Bradley reports no value for any probability or pseudoprobability of *a true negative* or *a false positive*, his data provide the basis for establishing exact values for them. The analysis which leads to these results starts with the calculation of the relative frequency of the certain event of *a non-defect*.

Bradley's assignment of his count of indications beyond 3 inches from a "defect" to *a false positive* means that there are countable data points in *a non-defect*. While there are data points, there are no physical objects. The relative frequency, however, is given by the ratio of physical objects to data points. It follows that the value of the relative frequency of the certain event of *a non-defect* is 0.

As the value of this relative frequency is not 1, any measure of a subset of *a non-defect* must be a pseudoprobability and not a probability. It follows that *a true negative* and *a false positive* have pseudoprobabilities. The values of these pseudo-probabilities sum to 0.

One knows from the definition of a *measure* (Halmos, 1950) that the two pseudoprobabilities have values greater than or equal to 0. It follows that the pseudoprobabilities of *a true negative* and *a false positive* have the value of 0.

While Bradley dismisses the absence of any probability of *a true negative* or *a false positive* from his report as unimportant in a safety study, this expresses the opinion that there is absolutely no value in keeping nuclear power generating equipment operating, for it is well known that any test's probability of *a true positive* may be improved arbitrarily, at the expense of increasing its probability of *a false positive*. This can be accomplished, for example, by issuing indications at randomly chosen locations of a steam generator tube. A less value-laden explanation of the absence would be that the ASME-USNRC test is inconsistent with the definition of *probability*.

CONCLUSIONS AND RECOMMENDATIONS

The irregularities in Bradley's pseudopopulation may be represented by Figure 2's Venn diagram. In this representation, the two circles represent the "defects" which correspond to *a defect*. One of the two circles represents those "defects" which are assigned to *a true positive* by the ASME-USNRC test. The other circle represents those "defects" which are assigned to *a false negative* by the test. The area of the rectangle which is not covered by a circle represents the event of *a non-defect*. It is not covered by physical objects. The region of overlap between the two circles is nil when the level of damage in a steam generator tube is nil. When the level of damage becomes dangerous, the two circles overlap totally.

The "defects" of the ASME-USNRC test, then, form a "population" which combines complete lack of coverage with complete overlap. Yet this extremely irregular population has been the support for estimates of "probabilities" which have been published for the consumption of nuclear reactor engineers by the USNRC! This incident prompts our interest in where else this kind of confusion may lie hidden.

There is reason to believe that this is not an isolated instance of a severely irregular population. The potential for them lies in the persistent definition of NDE as a field which detects "defects" in materials (Weismantel, 1975). Under pressure to establish NDE's reliability, NDE's scientists have taken these "defects" to occupy the populations of their studies. For example, the authors of *The Reliability of Non-destructive Inspection* have stated only that "defects" occupy the populations (Silk et al., 1987). Others have offered the equivalent view that flaws occupy the populations. The USNRC's research director has joined a former president of the ASME in stating this position (Beckjord, 1991; Fernandes, 1994). A scientific advisor to the USNRC on the reliability of NDE in nuclear reactors has offered this same view (Bush, 1991). So have the authors of a study of NDE's reliability in a British nuclear reactor (Cartwright et al., 1988). However, these views have definitely been wrong with respect to the kind of test which assigns "defects" to data points by the proximity of "defects" to indications. This has been the usual kind of test in the non-destructive examination of nuclear reactor components. The ASME-USNRC test of nuclear steam generator tubes provides an example.

Each such test assigns a "defect" to *a false negative* when it assigns it to *a true positive* or assigns no physical object to *a non-defect*. The degree of under coverage of *a non-defect* may be decreased by a loosening the rule of evidence which binds an indication to a "defect," but this increases the degree of overlap between the sets of physical objects corresponding to *a true positive* and *a false negative*. Whether there is under coverage, overlap or both, the "populations" of these "defects" are pseudopopulations.

The architects of tests of this kind might have responded to calls for establishing NDE's reliability by abandoning these tests or by writing the reports of studies of their reliability in the language of pseudoprobabilistics. However, they have responded in neither way. Instead, they have mixed the incompatible concepts of probabilistics and pseudoprobabilistics.

The result of this confusion can be dangerously misleading. Thus, we recommend its elimination by the translation of the reports of past studies in pseudoprobabilistics into the language of pseudoprobabilistics. The translation should be accomplished quickly, in view of the obvious dangers of continued confusion.

For the future, pseudoprobabilistic studies should be clearly discriminated from the probabilistic ones by the language of their reporting. However, there would be a problem in stopping with this reform only. Pseudoprobabilistics is not a substitute for probabilistics because probability measures risk but pseudoprobability does not. When the value for the pseudoprobability of a certain event falls below 1, this signifies a lack of evidence for the establishment of a level of risk. When the value for the pseudoprobability of a certain event falls above 1, this signifies ambiguity about the level of risk. Only probability measures the risks of NDE.

ACKNOWLEDGEMENTS

It is a pleasure to recognize Jean Perdijon of France's Commissariat A L' Energie Atomique for engaging in a correspondence that illuminated the paper's theme. Barbara Von Haunalter was gracious enough to donate her time to the preparation of graphic art.

REFERENCES

Beckjord, E., Letter to Terry Oldberg, Jan. 9, 1992.

Bowker, A. H. and G. J. Lieberman, *Engineering Statistics*, Second Edition, 1972. Prentice-Hall, Englewood Cliffs, NJ.

Bush, S., Letter to L. Shao, U. S. Nuclear Regulatory Commission, Feb. 18, 1992.

Bradley, E. R., P. G. Doctor, R. H. Ferris and J. A. Buchanan, "Steam Generator Group Project. Task 13 Final Report: Nondestructive Examination Validation.," NUREG/CR-5185, 1988. U. S. Nuclear Regulatory Commission, Washington, DC.

Cartwright, D. K. and K. S. Leyland, "The Validation of Sizewell 'B' Ultrasonic Inspections at the Inspection Validation Laboratory, Risley Laboatory," *Non-Destructive Testing*, 1988, p 57, 1988.

Cochran, W. G., *Sampling Techniques*, Third Edition, 1977. John Wiley & Sons, Inc, New York, NY.

Fernandes, J., Letter to Terry Oldberg, March 16, 1994.

Halmos, P., *Measure Theory*, 1950. Springer-Verlag New York Inc., New York, NY.

Juran, J. M., *Quality Control Handbook*, 1974, p 23-1. McGraw-Hill Book Company, New York, NY.

Mayants, L., *The Enigma of Probability and Physics*, 1984. D. Reidel Publishing Co., Dordrecht, Netherlands; c/o Kluwer Academic Publishers, Norwell, MA.

Perdijon, J., "Specification and Acceptance in Nondestructive Testing," *Materials Evaluation*, July 1993, pp 805-807.

Popper, K. R., *The Logic of Scientific Discovery*, 1959. Basic Books, Inc., New York, NY.

Silk, M. G., A. M. Stoneham and J. A. G. Temple, *The Reliability of Non-destructive Inspection*, 1987, p 36. Adam Hilger, Bristol, UK.

Weismantel, E. E., "Glossary of Terms Frequently Used in Nondestructive Testing," *Materials Evaluation*, Vol. 33, No. 4, Apr. 1975, pp 23A-46A.

FABRICATION OF SPECIMENS WITH CONTROLLED FLAWS

Robert L. Edwards and George J. Gruber
Nondestructive Evaluation Science and Technology Division
Southwest Research Institute
San Antonio, Texas

Paul D. Watson
Structural Systems and Technology Division
Southwest Research Institute
San Antonio, Texas

ABSTRACT

Most nondestructive evaluation (NDE) codes and standards require that the NDE equipment be calibrated using a calibration block. Ultrasonic testing (UT) historically has required the use of side-drilled or flat-bottom holes or notches. Recent technology has recognized that the acoustic response of "real" flaws is not directly comparable to artificial reflectors. The need arose to manufacture UT test specimens that contained real flaws of known size, shape, position, and orientation. The 1989 Section XI ASME Code, Appendix VIII (ASME Code,1989), requires NDE qualification of equipment, procedures, and personnel utilizing full-scale test specimens with actual (real) flaws. The same technology could prove of great benefit to industries other than nuclear, particularly for the fracture mechanics approach to fitness-for-purpose or lifetime-extension programs. This paper describes an approach to the design and fabrication of NDE test specimens with controlled flaws.

INTRODUCTION

Recent application of fracture mechanics approaches to establish fitness-for-purpose criteria that would extend (or establish) the life of pressure vessels, piping systems, and structural components has dictated the use of nondestructive evaluation (NDE) techniques to not only detect a flaw but to accurately size it as well. This has prompted many owners to acquire suitable NDE specimens to demonstrate the performance of the various NDE methods, procedures, and examiners for the required tasks. The historical use of artificial reflectors such as side-drilled and flat-bottom holes or notches for the ultrasonic testing (UT) method is being replaced by use of actual fabrication or service-induced flaws which display an acoustic response more typical of those encountered in practice. For instance, the newly invoked Appendix VIII (ASME Code, 1989) requires qualification of equipment, procedures, and personnel utilizing full-scale test specimens with actual flaws. This has increased the need for flaws that can be produced and accurately controlled in various test specimens. The typical service-induced flaws may consist of one of the following, depending on the material type and service conditions: mechanical fatigue; thermal fatigue; stress-corrosion cracking; and weld-induced flaws (during manufacture) such as slag, porosity, incomplete fusion, and weld or heat-affected-zone cracks.

Although the above-mentioned regulations are presently confined to the nuclear industry, the NDE technology can be applied to a host of other critical components in other industries such as fossil-fuel plants, chemical plants, oil and gas production and distribution systems, bridges and buildings, offshore platforms, airframes, and many other systems.

As the requirements and need for full-scale mockups and specimens duplicating the geometry and material with actual flaws increased, work also has increased to develop more cost-effective techniques to produce actual flaws. Since most service-induced flaws are generated over long periods of time and under adverse conditions, they are difficult and expensive to duplicate in a small test specimen or component segment such as a reactor pressure vessel (RPV) nozzle. Another difficult task in producing flaws in test specimens is controlling their location and size. Unique techniques have been developed for inducing cracks or other flaws that circumvent these problems and can be utilized for the fabrication of representative NDE specimens for personnel training, procedure development, and qualification of NDE techniques for any of the above-mentioned systems.

The typical NDE standards require that induced flaws have accurately controlled size, position, and orientation. Control of the size includes length, height (throughwall dimension), and width (particularly the width of a crack tip or slag inclusion). The position of the flaw(s) within the specimen must be known to adequately demonstrate a given technique, procedure, or operator's capabilities. Position includes all three coordinates as measured from a set point.

The following sections discuss the approach for fabrication of these critical specimens.

FABRICATION-INDUCED FLAWS

Welding-induced flaws experienced during fabrication of a component include slag, incomplete fusion (IF), inadequate penetration (IP), excessive porosity, and weld or heat-affected-zone (HAZ) cracking. All of these flaws can be intentionally induced into an NDE test specimen; however, the control of size, position, and orientation to the accuracy required is another matter. Side-wall IF and root IP are easily simulated and controlled. Actual porosity is easily induced, but to control the number and size of pores is very difficult. Fortunately, porosity is a fairly innocuous welding flaw and is seldom sized by UT techniques.

A technique has been developed that induces slag flaws of known size and shape. The slag flaw can be formed as single or multiple cavities, as illustrated in Figure 1. This is achieved by stacking cavities horizontally with each subsequent weld pass, or the individual slag cavity can be separated in the length direction. The slag pocket can be placed against the weld preparation cavity sidewall, which forms a flat zone on the sidewall side of the slag cavity. The topping weld pass above the top slag cavity can be deposited in such a way that an IF flaw is generated at the top of the slag pocket. The resultant flaw is identical to naturally occurring slag in both character and ultrasonic response, as shown later in this paper. The boundary envelope around the flaw can be very accurately controlled.

Fabrication-induced cracking is the most difficult flaw to intentionally induce in a weldment with the control necessary for NDE standards. The primary cracking modes in components are weld-metal solidification (WS) cracks and hydrogen-induced cracking (HIC). The former is caused by weld contamination, inappropriate weld-metal chemistry, excessive restraint, or a combination of these factors. The latter, caused by the introduction of hydrogen into a susceptible microstructure, can manifest itself as HAZ or weld-metal cracking. Techniques and procedures have been developed to induce a WS crack in carbon and low-allow steels of highly controlled size, shape, location, and orientation. These flaws, induced directly into the component, do not involve generation of a crack in a separate specimen that is subsequently sized, shaped, and implanted into the NDE test specimen. They are identical to naturally occurring WS cracks and are acoustically similar to both HIC and service-induced cracks, as discussed in the next section.

SERVICE-INDUCED FLAWS

Service-induced and/or propagated flaws are the most dangerous of all in that they can lead to failure of the component. The major offenders are mechanical fatigue cracks (MF), thermal fatigue cracks (TF), and stress corrosion cracks, both intergranular (IGSC) and transgranular (TGSC). Detection and sizing are

5787   5X

FIGURE 1. TYPICAL IRREGULARLY SHAPED, STACKED SLAG CAVITIES, GENERALLY CYLINDRICAL IN SHAPE, WITH AN LOF FLAW AT THE TOP SLAG POCKET

of paramount importance and precipitated the requirements of Appendix VIII, Section XI, of the Code. The following briefly describes SwRI's approach to simulating these flaws.

FERRITIC WELD-SOLIDIFICATION CRACKS

Supplement 3 of Appendix VIII (ASME Code, 1989) to Section XI of the ASME Code for ferritic pipe specimens states that the test flaws shall be either thermal or mechanical fatigue cracks, the implication being that both types are similar in carbon steel. However, the physical character of any fatigue crack is dependent on the loads involved, the cyclic rate at which it is formed and propagated, and any other outside stresses that may be prevalent. For instance, a tight high-cycle mechanical fatigue crack is generally planar in nature with a relatively smooth face. Figure 2 shows a typical high-cycle mechanical fatigue crack, which displays a very narrow crack opening [i.e., approximately 0.0025 mm (0.0001 inch) or less along its full face].

On the other hand, low-cycle, high-load mechanical fatigue cracks can produce much rougher crack faces with branching and some secondary cracking. This is particularly true for cracks formed in complex stress fields involving biaxial or triaxial stresses. Thus, a service-induced mechanical or thermal fatigue crack might display rough or smooth crack faces, with or without secondary or branch cracking.

The WS technique produces cracks in ferritic steel that are reasonably planar with crack tips less than 0.020 mm (0.0008 inch) in width. Figure 3 shows a typical WS crack in A-533 base

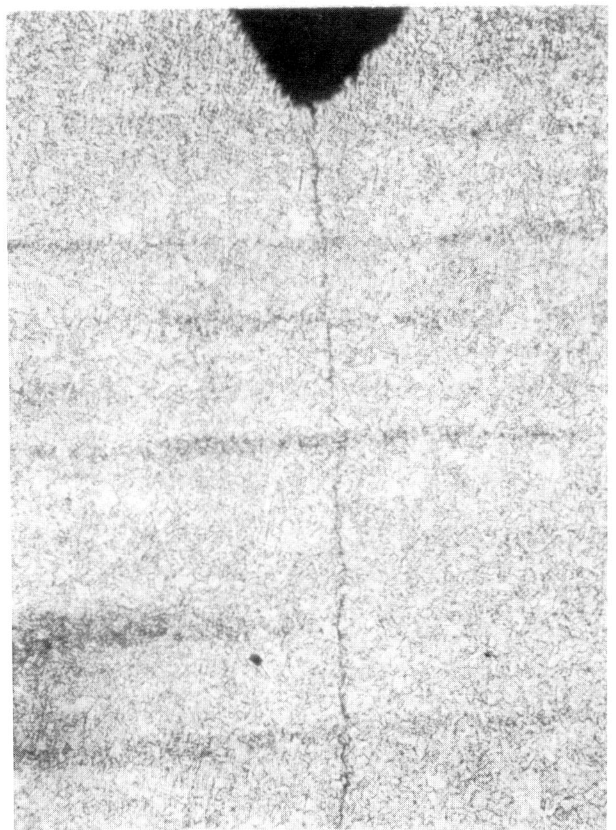

FIGURE 2. TYPICAL HIGH-CYCLE MECHANICAL FATIGUE CRACK. NOTE THE TIGHT AND PLANAR NATURE OF THIS CRACK.

material. This technique greatly reduces the amount of weld metal necessary to form a crack compared to the traditional approach for producing a mechanical or thermal fatigue crack. That is, implanting a coupon containing the crack into a cavity formed in the material.

The WS crack technique can be utilized to generate cracks directly into ferritic pipe or vessel test specimens of any shape or size to satisfy the crack requirements for ferritic pipe Supplement 3 or vessel Supplements 4, 5, 6, and 7 of Appendix VIII (ASME Code, 1989). Further, this type of crack can be generated in any component, mockup, or geometry that allow access to a small gas-tungsten-arc (GTA) welding torch. Thus, complex configurations such as T-K-Y connections of offshore platforms can be intentionally flawed.

Figure 4(a) shows the small cavity required to induce a surface-breaking WS crack located on a nozzle inner radius, and Figure 4(b) shows the small amount of weldment required. Figure 4(c) shows the fracture face of a typical WS crack on a nozzle inner radius. Figure 5 shows a service-induced (TGSC) crack in a gas

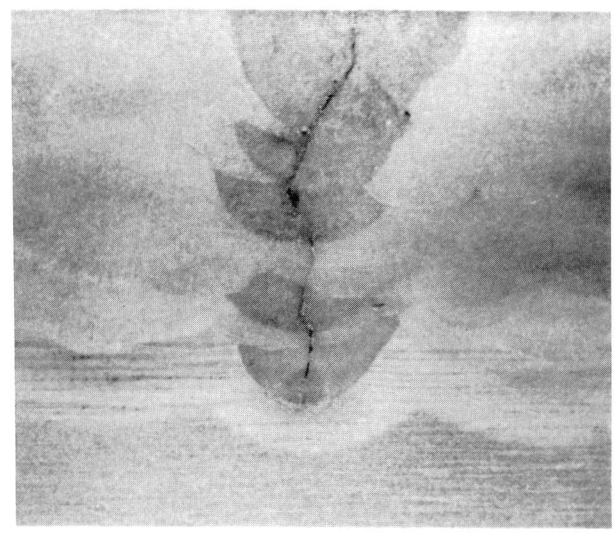

(a)

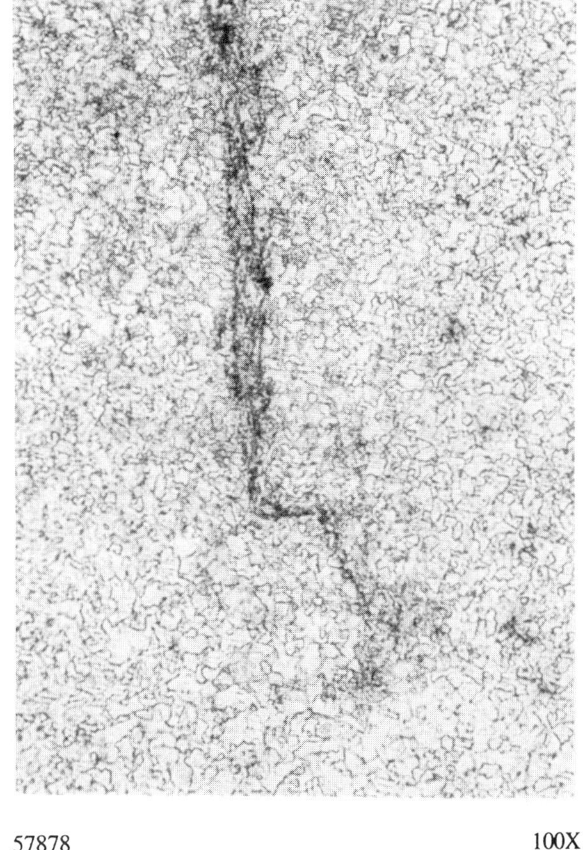

(b)

FIGURE 3. WELD-INDUCED CRACK IN FERRITIC WELD MATERIAL AT TWO DIFFERENT MAGNIFICATIONS

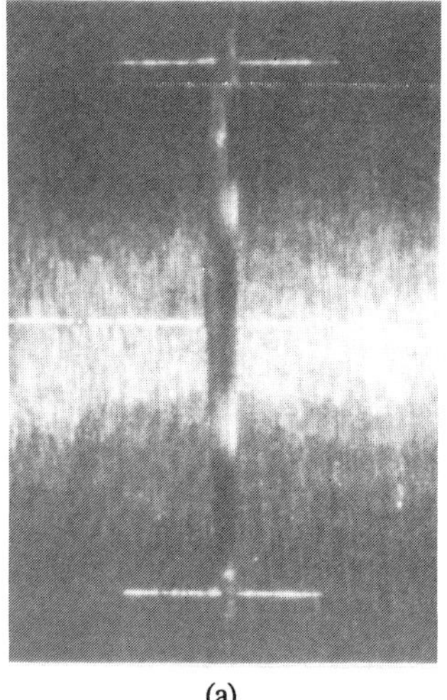

(a)

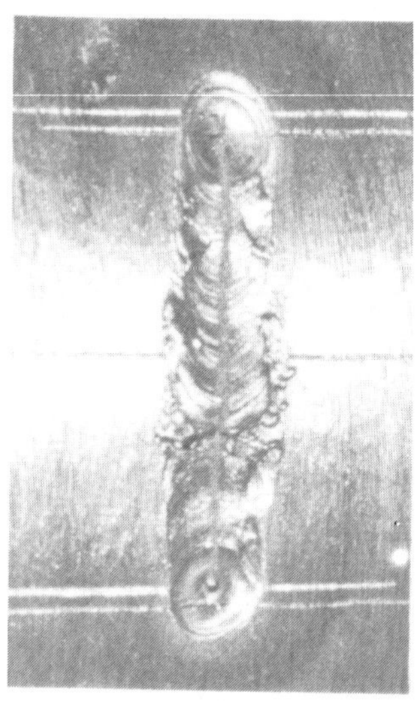

(b)

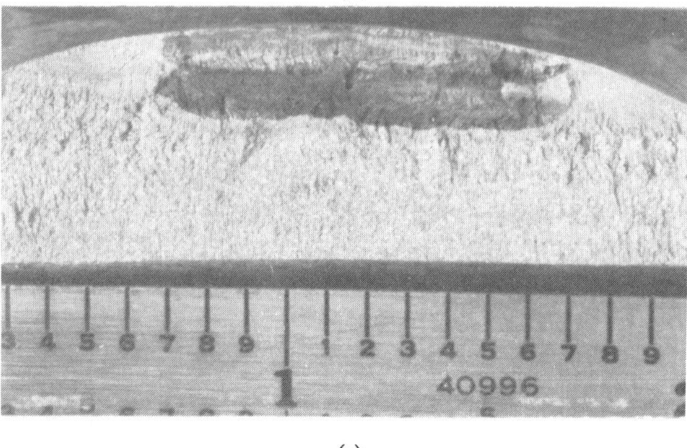

(c)

FIGURE 4. (A) SMALL CAVITY REQUIRED TO INDUCE A SURFACE-BREAKING WELD-SOLIDIFICATION CRACK LOCATED ON A NOZZLE INNER RADIUS, (B) THE SMALL AMOUNT OF WELDMENT REQUIRED, AND (C) THE FRACTURE FACE.

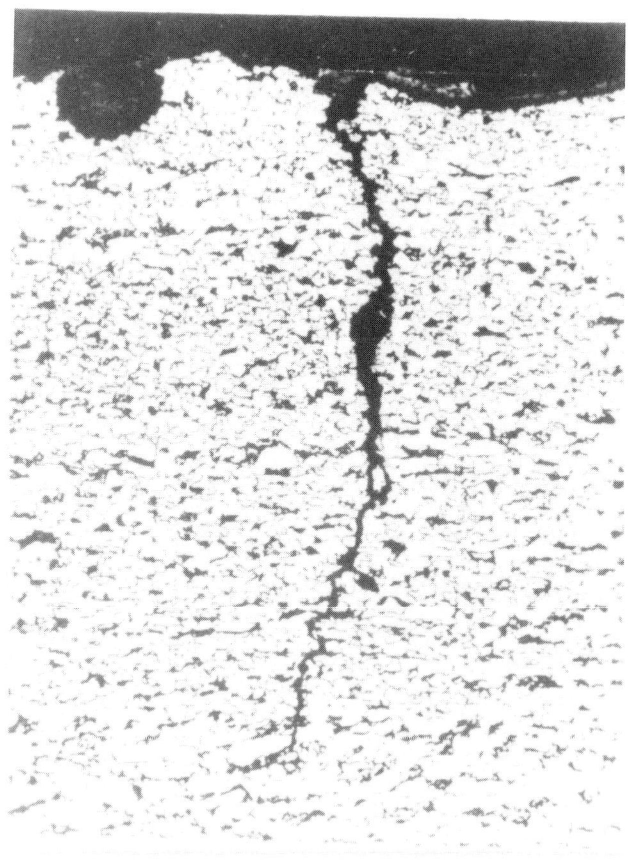

FIGURE 5. TRANSGRANULAR STRESS-CORROSION CRACKING (250X, NITAL ETCHED)

pipeline. Note that the character of this crack is nearly identical to the WS crack shown in Figure 3. These cracks can also be produced at any desired location and orientation by precise machining of the recessed cavities. Figure 6 shows a photograph of a 279-mm (11-inch) thick pressurized-water reactor (PWR) shell-to-nozzle mockup fabricated for NDE technique and procedure development using this process. (The ultrasonic responses of these and other crack types are shown later in this paper.)

AUSTENITIC WELD-SOLIDIFICATION CRACKS

Stainless Steel Material

The WS technique produces cracks in austenitic stainless steel material that more closely simulate IGSC cracks than do thermal fatigue cracks. The technique requires machining a narrow cavity directly into the material in the location and orientation desired. This cavity is filled with a specially formulated admixture of two chemically different austenitic filler metals that, upon solidification, produce a centerline crack in the weld deposit. The crack propagates from the weld-root fusion line to the surface. The crack is intercolumnar in nature and can have some branch cracking and secondary cracking; however, the main body of the crack

FIGURE 6. PHOTOGRAPH OF A 279.4-MM (11-INCH) THICK PWR SHELL-TO-NOZZLE MOCKUP FABRICATED FOR TECHNIQUE AND PROCEDURE DEVELOPMENT AS WELL AS PERSONNEL TRAINING

is reasonably planar. Figure 7(a) is a photomacrograph of a large [19-mm (0.750-inch)] WS crack in wrought Type 304 stainless steel which illustrates some branching and secondary cracking. Figure 7(b) is a photomacrograph of a small (12-percent-deep) WS crack in a 25.4-mm (1-inch) thick Type 304 stainless steel. Note the crack branching near the root of the weld. Figure 8 is a higher magnification of the branched tip portion of the crack. The width of the crack varies along its throughwall dimension. While the extreme tip of the crack is 0.0025 mm (0.0001 inch) wide, most of the crack tip has a 0.013-mm (0.0005-inch) width. The majority of the rest of the crack is 0.025 mm (0.001 inch) with a maximum width of 0.076 mm (0.003 inch) near the base, or opening, of the crack.

Nickel-Base Material

A similar weld-solidification cracking technique has been developed for nickel-base alloys such as Inconel 600 or weld deposits such as Inconel 182 or 82. This technique also utilizes a specially formulated admixture to create a weldment crack. The cavity is ground or machined in the same manner. Figure 9 shows the

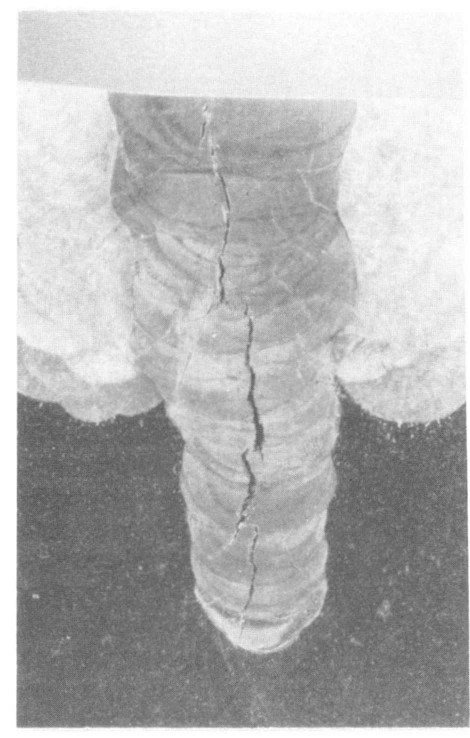

22910   3.8X

(a)

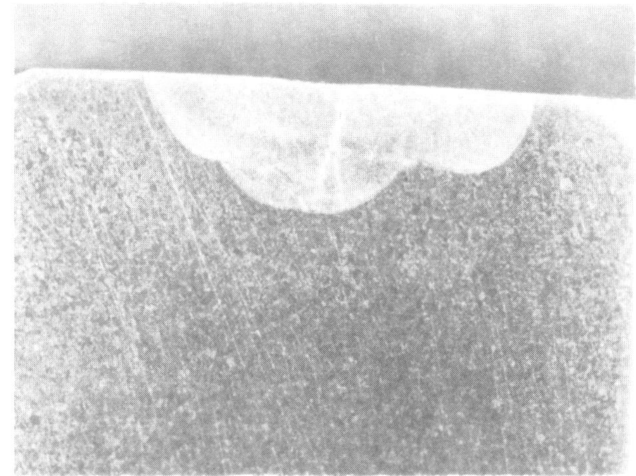

60577   5.25X

(b)

FIGURE 7. (A) PHOTOMACROGRAPH OF 19-MM (0.750-INCH) THROUGHWALL WELD-SOLIDIFICATION CRACK IN WROUGHT TYPE 304 STAINLESS STEEL AND (B) PHOTOMACROGRAPH OF A SMALL WELD-SOLIDIFICATION CRACK IN TYPE 304 STAINLESS STEEL. (SEE FIGURE 16 FOR ITS ULTRASONIC RESPONSE TO THE SwRI SLIC-40 TRANSDUCER.)

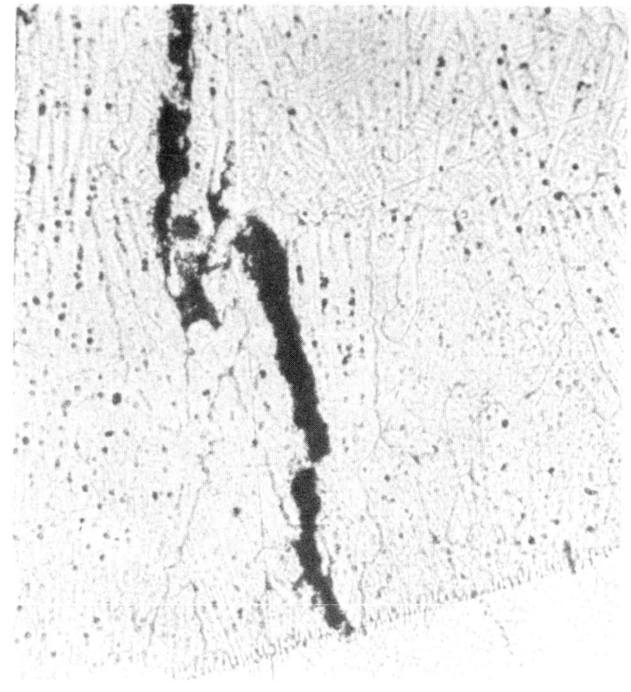

60578   200X

FIGURE 8. HIGHER MAGNIFICATION OF THE TIP OF THE RIGHT BRANCH. THE EXTREME TIP IS 0.0025 MM (0.0001 INCH) WIDE, BUT THE BULK OF THE CRACK HAS A 0.025-MM (0.001-INCH) MAXIMUM WIDTH.

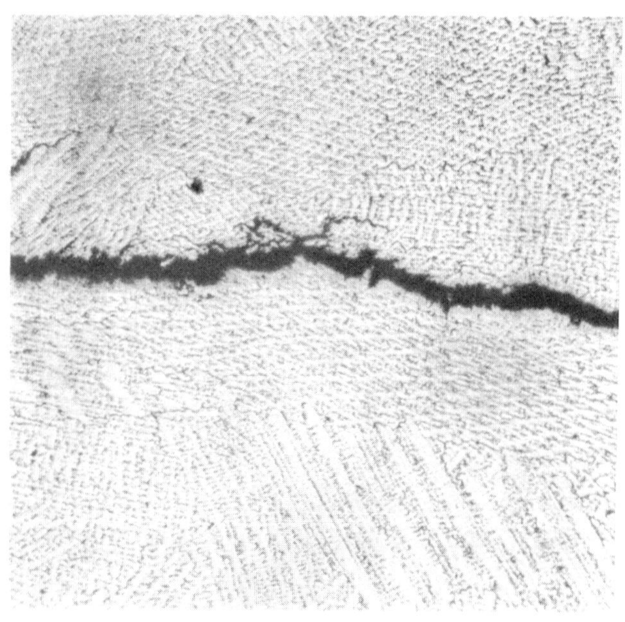

65860   50X

FIGURE 9. WS CRACK INDUCED IN INCONEL 600 MATERIAL, AS VIEWED ON THE SURFACE. NOTE THE VERY PLANAR NATURE OF THE CRACK.

nature of the crack on the surface of the deposit (top view) and is between 0.025 and 0.050 mm (0.001 and 0.002 inch) wide. Figure 10 shows the surface-connected portion of the crack in a cross-sectional view. Note the very small secondary crack adjacent to the main crack front.

Intergranular Stress Corrosion Cracks

Figures 11 and 12 show service-induced IGSC cracks taken from an operating nuclear steam supply piping system. The Figure 11 crack has a 0.559-mm (0.022-inch) width at the base and a crack tip width of less than 0.0025 mm (0.0001 inch). The Figure 12 crack is 0.127 mm (0.005 inch) wide near the base with the remainder varying from 0.041 mm (0.0016 inch) down to 0.0025 mm (0.0001 inch). Additionally, each of the cracks displays secondary cracking and branching. Note that the character of these flaws is not much different than the WS flaws in Figures 7 through 10.

Thermal Fatigue Crack

Figure 13 shows a thermal fatigue crack generated in a stainless steel pipe. The rough nature of the crack face and branch cracking are not significantly different than the WS crack shown in Figure 7(a).

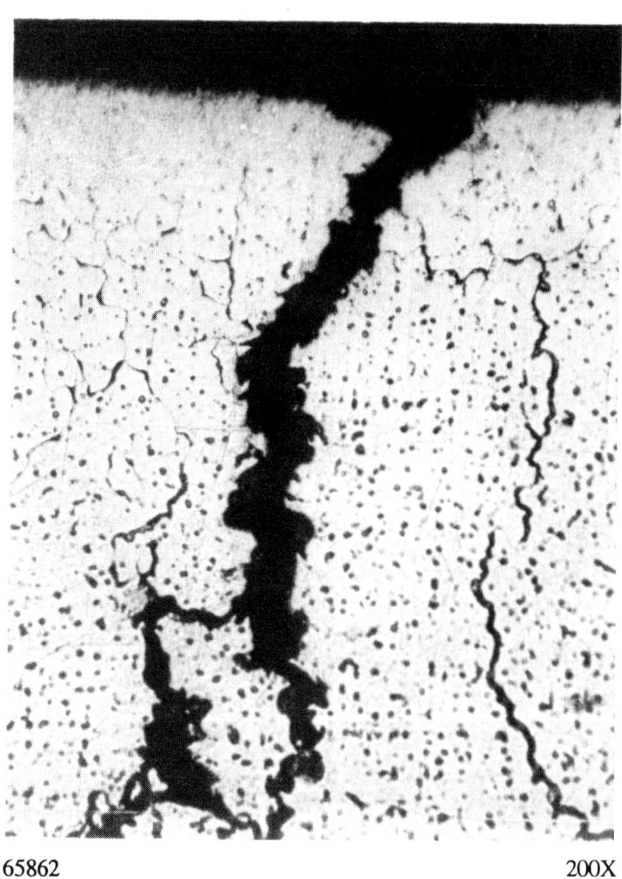

65862   200X

FIGURE 10. CROSS-SECTIONAL VIEW OF THE CRACK SHOWN IN FIGURE 9 AT HIGHER MAGNIFICATION

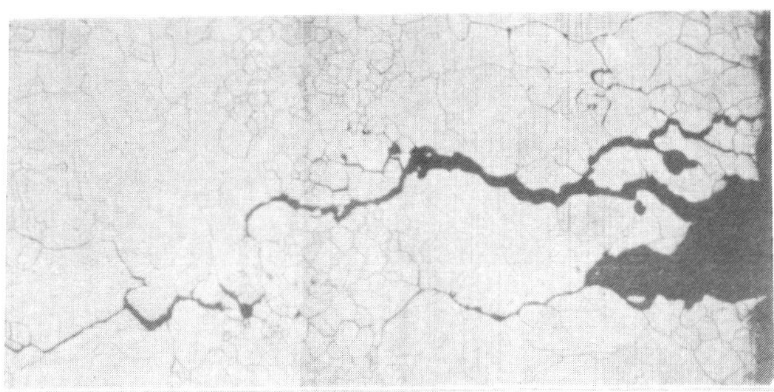

FIGURE 11. PHOTOMICROGRAPH OF AN IGSC CRACK REMOVED FROM A NUCLEAR STEAM SUPPLY SYSTEM PIPE. THE CRACK HAS A LARGE OPENING AT THE INSIDE SURFACE AND SECONDARY CRACKING.

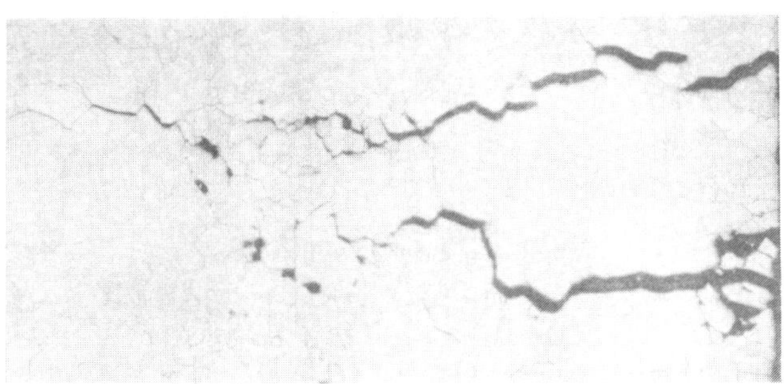

FIGURE 12. ANOTHER IGSC CRACK SHOWING TWO MAIN CRACKS AND RELATIVELY LARGE CRACK OPENINGS TO THE INSIDE SURFACE

The WS cracks are believed to be similar in physical character to IGSC and thermal fatigue cracks, with the major difference being an intercolumnar rather than intergranular nature. The ultrasonic response characteristics of the WS cracks are shown in the following section.

The WS flaws can be implanted in stainless steel castings, weld overlay, pipe-to-pipe welds, dissimilar weld joints, and Inconel safe ends and thermal sleeves.

ULTRASONIC EVALUATION OF THE FABRICATED CRACKS

RPV Test Specimens

NDE specimens have been fabricated in numerous configurations. The ferritic WS cracks and slag flaws have been evaluated and characterized. The ultrasonic responses of these flaws are suitable for specimens intended for procedure development, qualification, and personnel training. Figure 14(a) shows the single-tip ultrasonic response from a typical weld-solidification crack located approximately midwall of a PWR shell plate, and Figure 14(b) shows the multiple-tip ultrasonic response from a typical implanted slag flaw in a PWR shell plate.

FIGURE 13. THERMAL-FATIGUE WELD-ROOT CRACK INITIATING FROM A PARTIALLY REMAINING ELECTRO-DISCHARGE-MACHINED NOTCH (50X MAGNIFICATION)

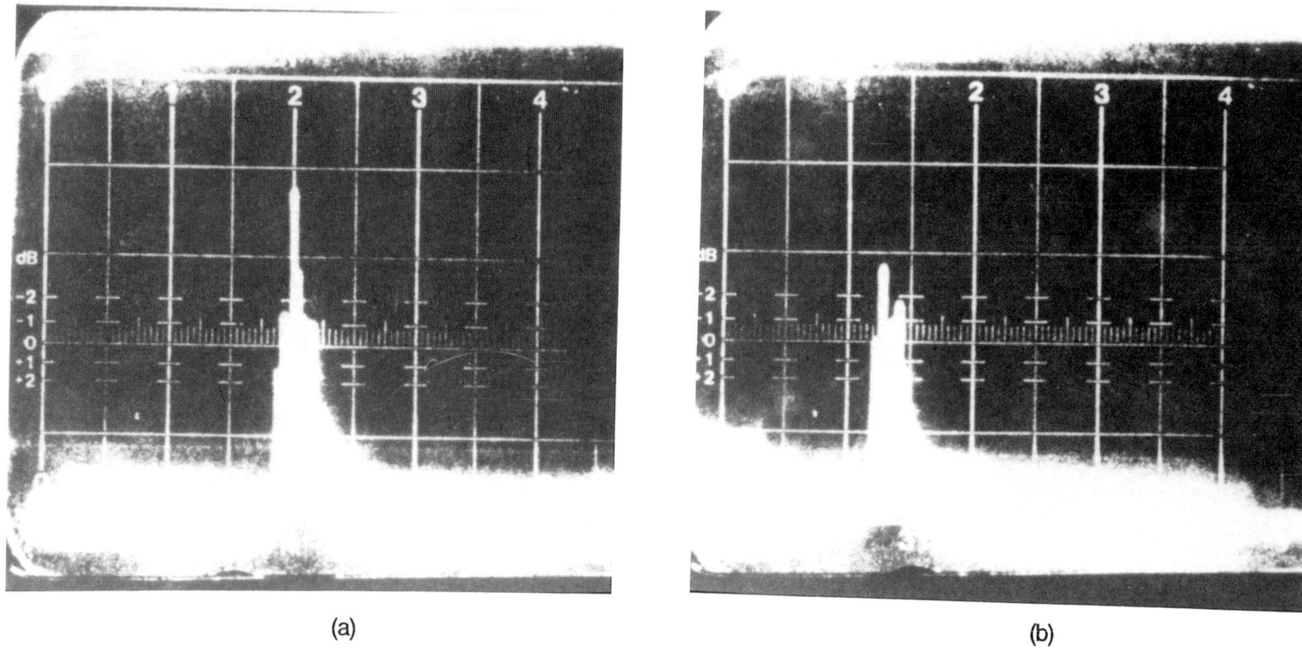

(a)   (b)

FIGURE 14. ULTRASONIC RESPONSES FROM (A) WELD-SOLIDIFICATION CRACK IMPLANTED APPROXIMATELY MIDWALL OF PWR SHELL PLATE AND (B) SLAG FLAW IMPLANTED APPROXIMATELY MIDWALL OF PWR SHELL PLATE

Austenitic Pipe Specimens

Ultrasonic response measurements were made on four crack types with a standard 2-MHz, 45-degree shear-wave (S45) probe and a special 4-MHz SLIC-40 transducer module (Gruber, et al., 1984; Gruber, 1986). Figure 15 shows the four pulses that result from the interaction of the SLIC-40 with an inside-surface-connected crack. The objectives of these measurements were two-fold: (1) to establish the similarity between the ultrasonic responses of IGSC cracks and WS cracks and (2) to demonstrate the large differences between the ultrasonic responses of IGSC cracks and thermal or mechanical fatigue cracks. The S45 crack-tip and crack-base signals are labeled pulses 1 and 2, respectively, in Figures 16, 17, and 18.

Comparison of IGSC Cracks with Weld-Solidification Cracks.

The difficulty of the S45 probe to detect a small (less than 20-percent-deep) IGSC crack in an austenitic pipe is well known. Figure 16(a) shows the weak base-reflected signal (pulse 2) received at midscreen from a small IGSC crack. The comparable weak pulse 2 obtained from the 12-percent-deep WS crack photographed in Figure 6(b) is shown in Figure 16(b).

All screen presentations were obtained with the same instrument gain setting. The small WS crack and the small IGSC crack appear to be about equally difficult to detect and size, resulting in a minor difference in amplitude response. This establishes the close similarity between WS and IGSC cracks.

Comparison of IGSC Cracks with Mechanical and Thermal Fatigue Cracks.

The difference in behavior between mechanical (and thermal) fatigue cracks versus IGSC cracks can be seen by comparing Figure 18 (for mechanical fatigue) and Figure 16 (for IGSC). The tip-diffracted signal (pulse 1) for an S45 probe is very prominent in Figures 17 and 18, while it is unobservable in the case of IGSC (Figure 16).

Similar results are shown in Figure 18 for a 25-percent-deep mechanical fatigue crack. It may, therefore, be concluded that thermal and mechanical fatigue cracks are not capable of simulating the weak ultrasonic signals typically received from IGSC cracks in austenitic pipe specimens.

SUMMARY

In summary, the presented results indicate that the WS cracks are closer, both metallurgically and ultrasonically, to IGSC cracks than to mechanical and thermal fatigue cracks. The ultrasonic response characteristics from thermal and mechanical fatigue are actually closer to those of surface-connected notches rather than IGSC cracks.

The results of these evaluations support the recommendation to use WS cracks for both the austenitic and ferritic specimens for personnel training and qualification and for procedure development and qualification.

CONCLUSIONS

Techniques and procedures have been developed to produce typical fabrication-induced and simulated service-induced flaws in a range of alloys. The procedures ensure adequate control of size, shape, location, orientation, and character so as to be applicable as NDE reflectors for UT examinations and suitable for

14

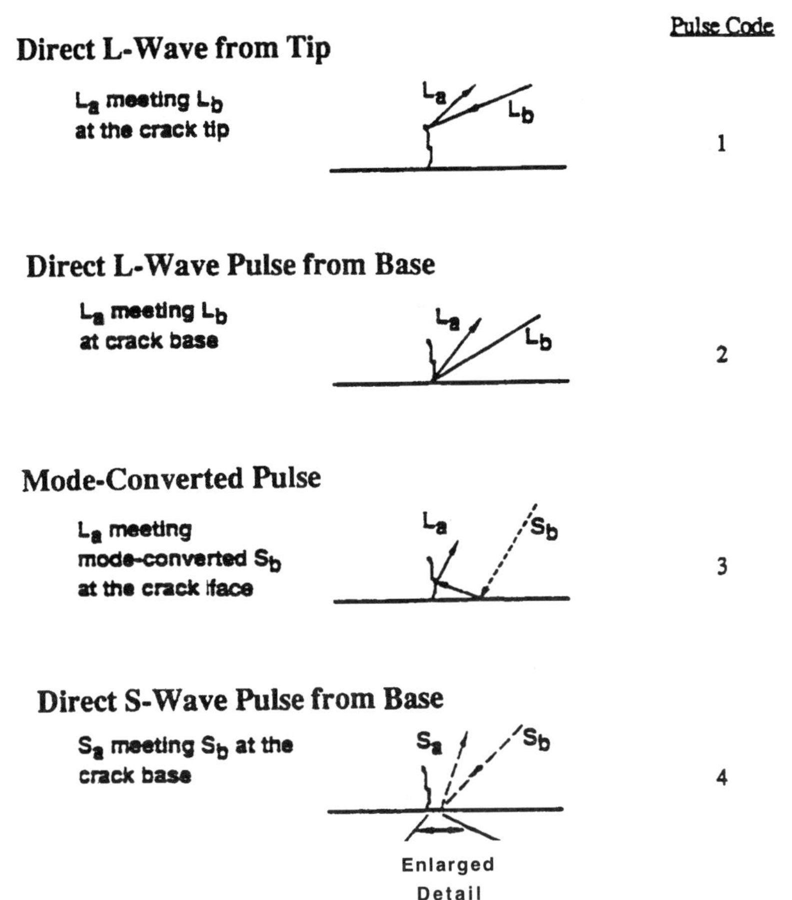

FIGURE 15. INTERACTION OF THE INCIDENT L-WAVE AND S-WAVE FROM THE SLIC-40 TRANSDUCER WITH AN INSIDE-SURFACE CRACK RESULTING IN FOUR ASSOCIATED PULSES

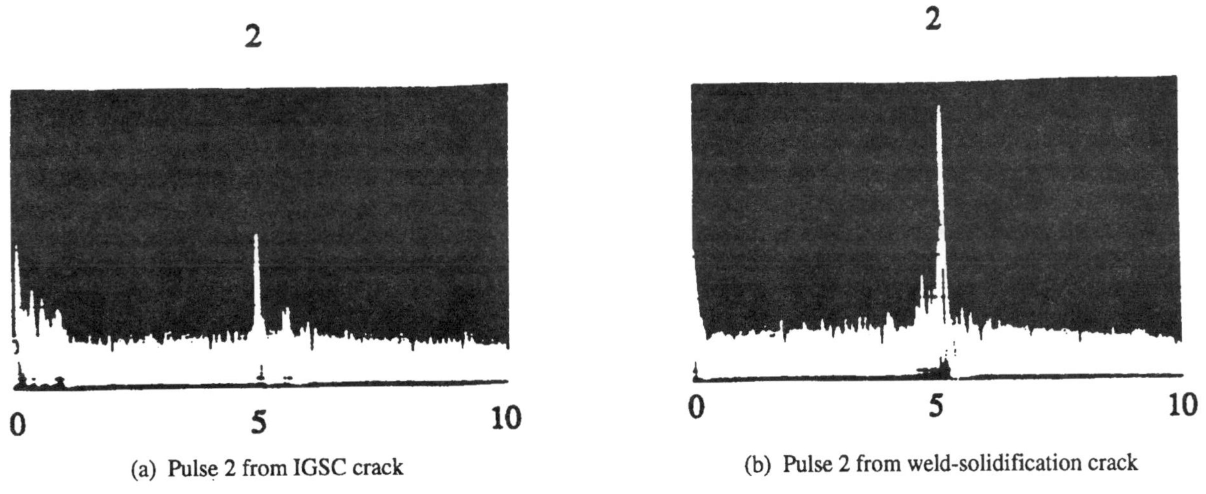

(a) Pulse 2 from IGSC crack

(b) Pulse 2 from weld-solidification crack

FIGURE 16. ILLUSTRATING THE DIFFICULTY OF THE S45 PROBE TO DETECT AND SIZE A SMALL IGSC CRACK (A) AND WELD-SOLIDIFICATION CRACK (B)

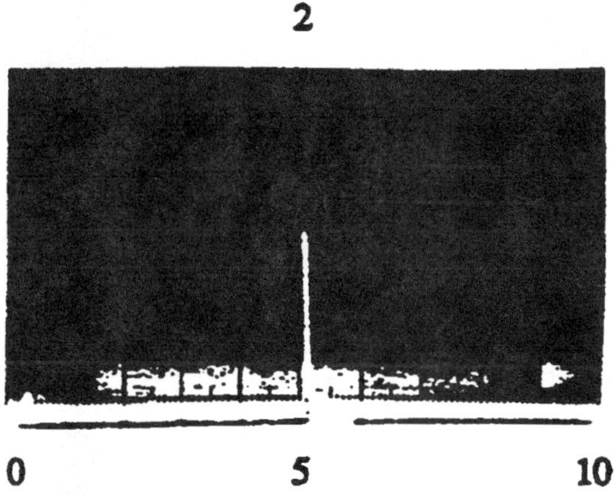

FIGURE 17. ILLUSTRATING THE EASE OF DETECTING A VERY SMALL THERMAL FATIGUE CRACK WITH THE S45 PROBE

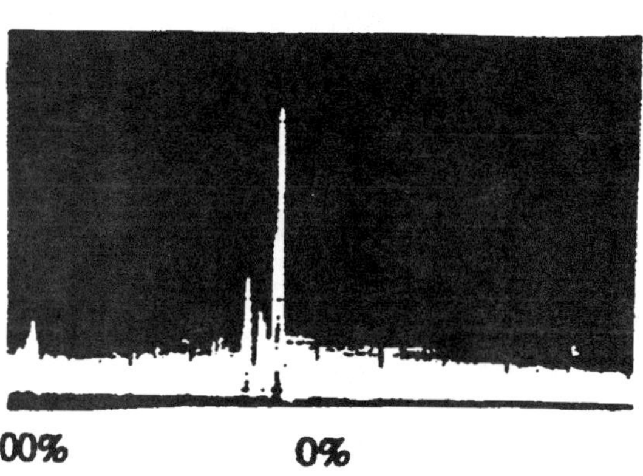

FIGURE 18. ILLUSTRATING THE EASE OF DETECTING AND SIZING A 25-PERCENT-DEEP MECHANICAL FATIGUE CRACK WITH THE S45 PROBE

radiographic, magnetic particle, and liquid penetrant examinations for procedure and equipment validation and personnel training and testing. Although originally developed to service the nuclear industry, this technology is directly applicable to other components that are considered to be critical enough to perform NDE operations in service. The catastrophic failure of a heavy-wall pressure vessel, chemical vessel, power piping, gas or oil pipeline, or offshore platform can have disastrous consequences.

REFERENCES

ASME Boiler and Pressure Vessel Code, Section XI, Appendix VIII, 1989 Edition.

Gruber, G. J., Hendrix, G. J., and Schick, W. R., 1984, "Characterization of Flaws in Piping Welds Using Satellite Pulses," *Materials Evaluation*, Vol. 42, pp. 426-432.

Gruber, G. J., 1986, "Monitoring of Intergranular Stress Corrosion Cracks in Pipes with Weld Overlays," *Int. J. Pres. Ves. and Piping*, Vol. 22, pp. 225-232.

NEAR AND FAR FIELD EXPERIMENTAL EVALUATIONS OF ULTRASONIC TRANSDUCERS

Bin Hu
Department of Mechanical Engineering
Tennessee Technological University
Cookeville, Tennessee

J. R. Houghton
Department of Mechanical Engineering
Tennessee Technological University
Cookeville, Tennessee

ABSTRACT

Five different flat and focused transducers with frequencies from 1 MHz to 20 MHz were calibrated. Results from the ASTM standard steel ball method plus a new pinducer technique are presented and discussed. The uniqueness of each transducer's near field is shown to be a hazardous region for measurement of flaw sizes that requires a high accuracy. The far field is confirmed to be uniform in behavior but this field also requires calibration checks of individual transducers. The focused zone of two focused transducers are discussed with respect to the NDE measurement of a flaw size in the beginning, middle and end of the zone.

INTRODUCTION

Materials testing and evaluation are very important for controlling and improving product quality in manufacturing processes. The ultrasonic method has become a powerful tool for non-destructive evaluation (NDE) of manufactured products. A major component of an ultrasonic testing system is the transducer. In other words, NDE results are very dependent on the piezoelectric element, which converts electrical signals into mechanical vibrations in the generation mode, and uses an inverse effect for detection of defects.

This paper discusses near field and far field characteristics of the individual flat and focused immersion piezoelectric transducers. Five different frequencies transducers were selected and used for the research reported in this paper..

The theoretically radiated field of all ultrasonic transducer for both continuous wave and transient excitation are compared generically to the experimental results in this paper. An on-axis profile of the radiated field and a three dimensional profile of pressure distribution of piezoelectric ultrasonic transducer, space-domain response (Papadakis, 1979), will be provided for both the theoretical and the experimental methods.

There are several different techniques for characterizing the immersion transducers. The most common approach is to use the immersion transducer in a pulse-echo mode by receiving the signal reflected from a small steel ball (ASTM Designation: E 1065-87 a). The test results will provide amplitude, spatial and spectral information. A second method is to use a flat fluid/solid interface to replace the point reflector as described by Bahr *et al.* (1985). A third technique is to use a scanned miniature hydrophone receiver in the fluid, as described by Lewin (1981). A fourth, less expensive method to characterize the transducer uses a pinducer instead of the hydrophone and is presented in this paper. Our pinducer was custom built for $100.00 by Valpey-Fisher Inc.. It has a small aperture of 0.02 inch diameter, requires no expensive high impedance pre amplification, and has a uniform response curve from 1 KHz to 7 MHz. Several laboratories have used the pinducer successfully for acoustic emission research.

A B-scan testing system from SONIX Inc. was used to evaluate the immersion transducer. The instrumentation system includes a Panametrics 5052 pulser/receiver, an ultrasonic scanning machine with 3-degrees of motion, a motor controlling box, an IBM compatible PC, a test transducer and a steel ball. The testing system is illustrated in Figure 1.

THEORY

Hutchins and Hayward (1990) and Zhang (1980) described the necessary relationships to predict the field of a transducer theoretically. The transducer is considered to be an ideal radiator known as a "plane piston" surrounded by an infinite baffle. The theoretical analysis results for a single frequency continuous-wave excitation is developed in Hutchins and Hayward's work (1990). A sample on-axis pressure amplitude distribution is illustrated in Figure 2. A sound field pressure profile along a diametral plane from the continuous-wave excitation is illustrated in Figure 3 for the transducer radius of 5λ, where λ is the radiated wavelength in water.

For continuous excitation, the on-axis pressure profile exhibits a series of maxima and minima amplitudes within the near field. The far field provides a smooth, uniform, but nonlinear value. The diametral plane profile plots in Figure 3 show the sound pressure distribution in front of the transducer. It is interesting to note the many wrinkles around the profile surface. There is a lot more surface roughness than one would expect from the values shown for the on-axis profile. The reason for interest in the 3-D pressure plot is to test the hypothesis that since the signal detected by an ultrasonic transducer is the reflection from acoustic pressure integrated across an area at a gated depth a flow measurement.

When a sine pulse signal is used to excite the piezoelectric element, theory shows a much smoother sound pressure feature. The on-axis and 3-D sound pressure amplitude plots for a circular plane piston with one cycle of a sinusoid (center frequency f=c/λ) are presented in Figure 4. In the on-axis figure, Figure 4 (a), a constant value is exhibited from the surface of transducer to some distance along z and a rise to a maxima amplitude called the near field followed by a slow fall is the region called the far field. The 3-dimension pressure plot shows a much smoother curve. This would be much better than Figure 3 for measurement purpose.

Work by Weyns(1980) indicates that the transient field profile will approach the continuous-wave situation as the excitation signal becomes a series of sine wave pulses. Figure 5 presents the on-axis and 3-D pressure profiles from a circular plane piston with six cycles of a sinusoidal pulse.

EXPERIMENTAL RESULTS

In nondestructive evaluation, the transient or impulsive type excitation is used for many applications. Experimental sound field pressure from ultrasonic transducers approach the theoretical transient excitation case described above. The pressure fluctuation in the near field of a transducer will have a detrimental effect on the accuracy of information gathered. The working region of an ultrasonic transducer for NDE is the far field. The purpose of the research was to determine the diametral plane pressure profile of different ultrasonic transducers undergoing transient excitation.

Flat Transducer

Five flat ultrasonic transducers selected for characteristics evaluation are 1 MHz 0.5 inch Dia., 2.25 MHz 0.25 inch Dia., 5 MHz 0.25 inch Dia., 10 MHz 0.25 inch Dia. and 20 MHz 0.125 inch Dia.. The 1 MHz and 5 MHz are displayed in the paper.

Steel Ball as Target. The most commonly used measurement technique to evaluate characteristics of ultrasonic transducers is the reflections from a steel ball. As shown in Figure 1, a steel ball was positioned at one side of the tank as a target. To plot a 3-D pressure distribution of a transducer experimentally, a large number of time-domain signals need to collected along the axial and radial directions. A set of time domain data (200 points) along the x direction, at a fixed distance z from transducer surface

was collected. The data was from a B-scan program. To get a spectrum of one data set along the x direction, an FFT of the time domain data was used. One frequency corresponding to the design natural frequency of the transducer was selected as the specified frequency in the FFT data for reducing the data. Then the spectrum information from all the points selected was reduced using MATLAB, and 3-D pressure field was plotted for the particular transducer.

Figure 6 gives a pulse excitation signal for the 1 MHz 0.5 inch flat transducer. Figure 7 is the on-axis pressure distribution using a 0.5 inch diameter steel ball as a target. A pressure profile of the 1 MHz transducer in both axial and radial coordinates is illustrated in Figure 8. Figures 9, 10 and 11 present a pulse excitation signal, on-axis pressure profile and a 3-D pressure field for the 5 MHz flat transducer using a steel ball.

Pinducer as Receiver. The pinducer was mounted inside the tank opposite the test transducer. Two test results from the 1 MHz and 5 MHz flat ultrasonic transducer evaluations are shown in Figures 12 to 17. Figures 12, 13 and 14 show a received signal, on-axis pressure profile and a 3-dimension sound field plot for the 1 MHz flat transducer using a 0.02 inch diameter pinducer. A received signal, on-axis pressure profile and a 3-D pressure field for the 5 MHz flat transducer using a pinducer are presented in Figure 15, 16 and 17.

Comparison of the on-axis profile from the steel ball test, Figure 7, to the pinducer test, Figure 13, shows many similarities, The comparison of the 5 MHz test results indicates a much rougher response from the pinducer. The pinducer time domain plots are noticeably different and cleaner than the steel ball plots. Comparison of Figure 13 to Figure 5(a) shows that the far field contour shape is more representative of the mathematical model. Notice that the two major sine wave pulses in Figure 12 are responsible for the small peak at z = 0.3 inch shown in Figure 13. Also notice that the steel ball time domain pulse is significantly different.

Comparison of the 5 MHz transducer time domain plots, Figures 9 and 15, show a significant variation. The pinducer shows a collection of four sine pulses that would explain the increased near field noise. The on-axis pressure profiles are very different between the two test methods. The basic core shape of the pressure from the pinducer, Figure 16, seems to be more representative of the theoretical prediction. From this preliminary work, the indications are that the pinducer measurements are a more reliable measurement of the transducer acoustic pressure profile than the ASTM steel ball test.

Focused Transducers

Two focused ultrasonic transducers were characterized using a steel ball as the target. The transducers were a 5 MHz 0.5 inch Dia. with a 1.46 inch focal length and 5 MHz 0.5 inch Dia. with a 3.15 inch focal length. A 0.5 inch steel ball was used in this experiment.

The test results of focused transducers are shown in Figures 18 to 23. Figures 18 and 21 are the pulse excitation signals for the

5 MHz focused transducer with a specified 1.46 inch and 3.15 inch focal length, respectively. Using the manufacturer's values the calculated beginning and end of the focal zone for the 1.46 inch focal length transducer are 1.32 inch and 2.01 inch. The length of focal zone is 0.7 inch. The beginning and end locations of focal zone are 2.38 and 4.55 inch for 3.15 focal length transducer. The length of the focal zone is 2.17 inch. A on-axis pressure profile and a 3-D pressure field for 1.46 inch and 3.15 inch focal length transducers using a steel ball are present in Figures 19, 20, 22 and 23 respectively.

The characteristics evaluation process for focused transducers was the same as for the flat transducer. But the testing results on the axial pressure profile obviously are different. Instead of a rough near field and uniform far field regions, a Gauss probability shaped curve is shown as the on-axis pressure profile. As shown in Figures 19 and 22, the focal zone amplitude increases from 50 percent before the focal point. And decreases after the focal point to the 50 percent of peak value.

Transverse Focal Zone of Focused Transducer. Two focused transducers were evaluated in the focal zone to demonstrate the variations that could influence measurements and NDE conclusions. Three transverse profiles at the focal point, at the beginning and at the end of the focal zone of the 5 MHZ focused transducer with 1.46 inch and 3.15 inch focal length are presented in Figures 24 and 25. For the 1.46 inch focused transducer the experimental focal point was 1.34 inch. The beginning location was 0.94 inch and the end of the focal zone was 1.73 inch. The actual length of the focal zone was 0.79 inch. For the 3.15 inch focused transducer the experimental focal point was 3.54 inch. The beginning location was 2.52 inch and the end of the focal zone was 4.72 inch. The actual length of the focal zone was 2.20 inch. The test results indicate the nonuniform response that can exist in the focal zone. The authors think that this is a significant point that should be in the NDE engineer's mind when the C-scan specifications are made and the focal point location is specified. Signal amplitude in reflected energy from flaws of the same size will be fifty percent small at the beginning and end of the focal zone.

DISCUSSION AND SUMMARY

Flaw growth measurements by the authors in carbon composites structures have shown significant variation in the flaw size when either the ultrasonic transducers are changed or the pulser settings are changed. In one case, a delamination flaw was reported to have shrunk after additional fatigue damage. The flat and focused transducer pressure profile studies reported here have confirmed that there can be significant variations in the 3-D pressure around a flaw and that one must know the characteristics of each transducer and match the profiles before swapping transducers. If this precaution is not practices in NDE the results of C-scan comparison for a structure could be highly variable.

Both theory and experiment have demonstrated that the type of pulse excitation signal will effect evaluation results in the near field. In the steel ball test, the variation of the excitation signal does not cause a significant change in the near field. A different frequency signal does cause roughness and larger variations in the far field of the on-axis pressure profile. The pinducer test shows more roughness in the near field than that of the steel ball test but the pinducer is believed to be a better representation of the true pressure profile.

Previous comparisons of a steel ball reflection test to a hydrophone technique have shown the hydrophone to be more accurate results for transducer characteristics evaluation. But a hydrophone is very expensive. A pinducer is shown in this paper to be an inexpensive tool with results similar to a hydrophone.

The focal zone of a focused transducer is usually expected to be a uniform region of constant amplitude and radius for NDE. But, the widths of amplitude beam, shown in Figure 24 and 25, are not uniform at three location of the focal zone, and also vary from transducer to transducer. That will cause the accuracy of the sizes of flaws. The test results of the focused transducers from this experiment demonstrate this conclusion.

To do flaw evaluations using a C scan search unit, knowledge of the features of a transducer used are important. Each individual flat transducer can cause different test result in the near field. For all flaw location and NDE evaluation work it is highly recommended that the flat transducer be mounted to measure only in the far field. A condition for accuracy of measurement with a focused transducer is to know the pressure distribution and diameter variations of the focus zone.

ACKNOWLEDGMENT

An FFT data transformation program which was developed by Peitao Shen and written by Volker Klumpp has been used in this research. The authors are most grateful to Dr. Peitao Shen and Mr. Volker Klumpp for the use of their program.

REFERENCES

Bahr, R. K., Bucklew, J. A., and Flax, S. W. 1985. "Optimal Center-Frequency Estimation for Back-Scattered Ultrasound Pulses". A Trans Sonics and Ultrasonics. SU-32, pp. 809-813.

Hutchins, D. A., and Hayward, G., 1990, "Radiated Fields of Ultrasonic Transducers," Ultrasonic Measurement Methods, R. N. Thurston and Allan D. Pierce ed., ACADEMIC PRESS, INC., San Diego, Physical Acoustics Vol. XIX, pp. 1-79.

Lewin, P. A. 1981. "Miniature Piezoelectric Polymer Ultrasonic Hydrophone Probes". Ultrasonics **19**, pp. 213-216.

Papadakis, E. P. , 1979, "Theoretical and Experimental Methods to Evaluate Ultrasonic transducers for Inspection and Diagnostic Applications", A Trans. on Sonics and Ultrasonics, Vol. SU-26, pp. 14-27.

Zhang, Shouyu, 1980, "Ultrasonic Evaluation theory and ap applications." Mechanics industrial publications, Beijing, pp. 29-39.

Weyns, A. 1980."Radiation Field Calculations of Pulsed Ultrasonic Transducers". Ultrasonics **18**, pp. 183-189.

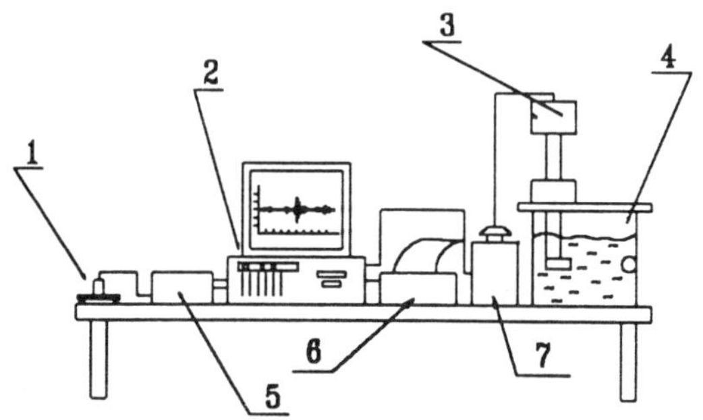

1. Ultrasonic transducer
2. PC-386 computer with C-scan
   and attenuation calculation program
3. X,Y,Z-axis Manipulator
4. Immersion tank
5. Panametrics 5052PRX50 Pulser/Receiver
6. HP Paintjet XL color printer
7. Motor controller

FIGURE 1: EXPERIMENT SETUP

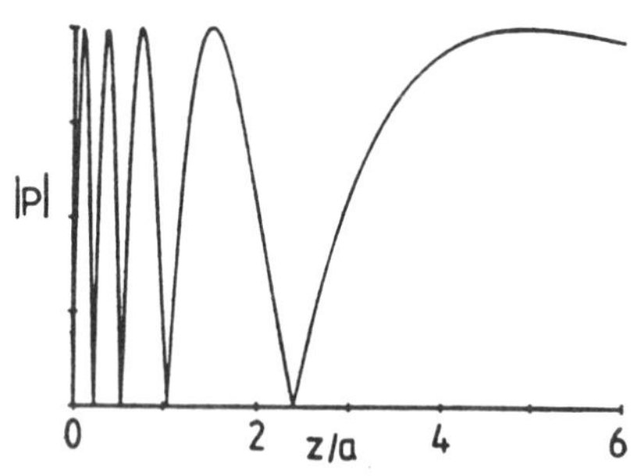

FIGURE 2: THEORETICAL ON-AXIS PRESSURE AMPLITUDE

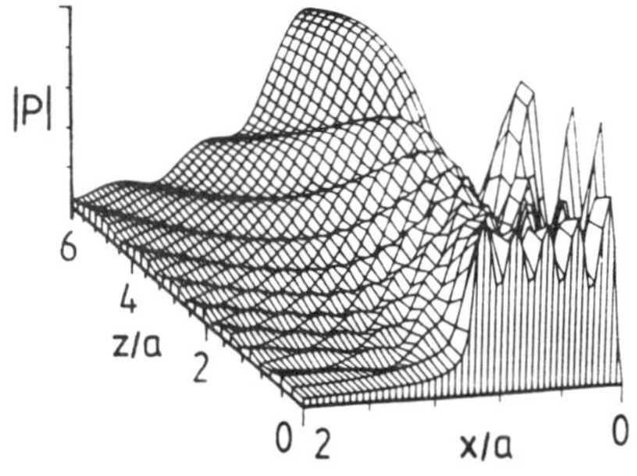

FIGURE 3: RADIAL PRESSURE PROFILE FROM CW EXCITATION, HUTCHINS AND HAYWARD(1990)

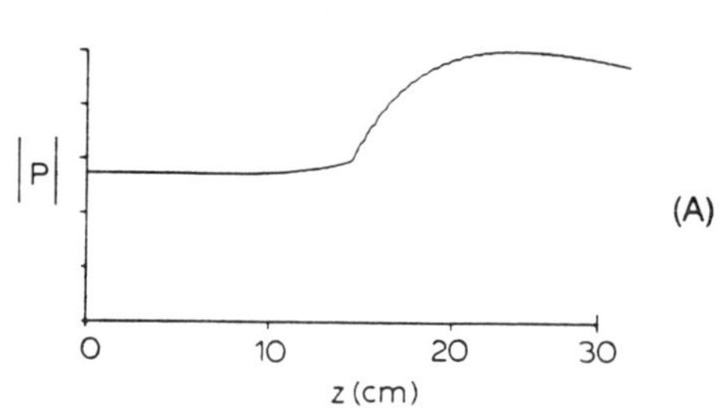

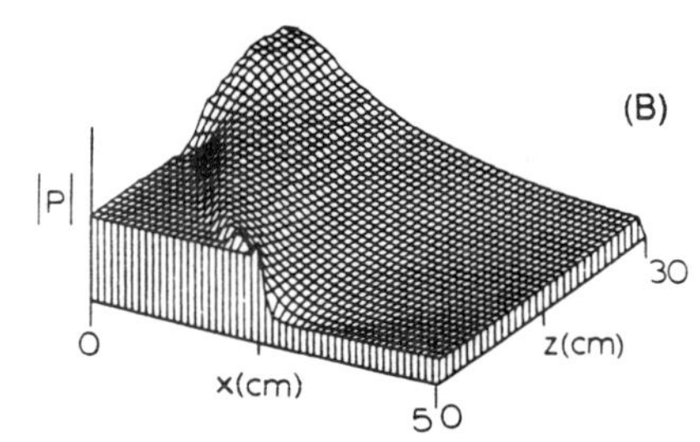

FIGURE 4: SIMULATED PRESSURE PROFILE WITH ONE CYCLE OF A SINUSOID. (A) AXIAL FIELD (B) FULL SECTION, WEYNS(1980)

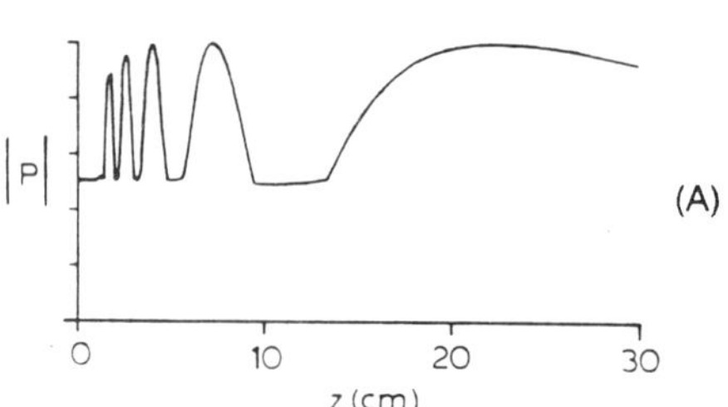

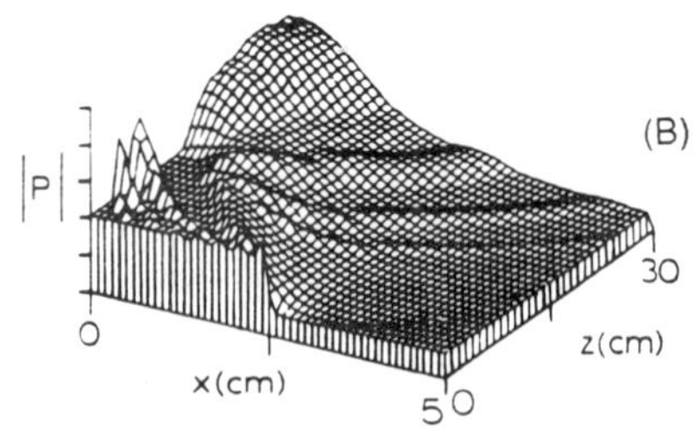

FIGURE 5: SIMULATED PRESSURE PROFILE WITH SIX CYCLE OF A SINUSOID. (A) AXIAL FIELD (B) FULL SECTION, WEYNS(1980)

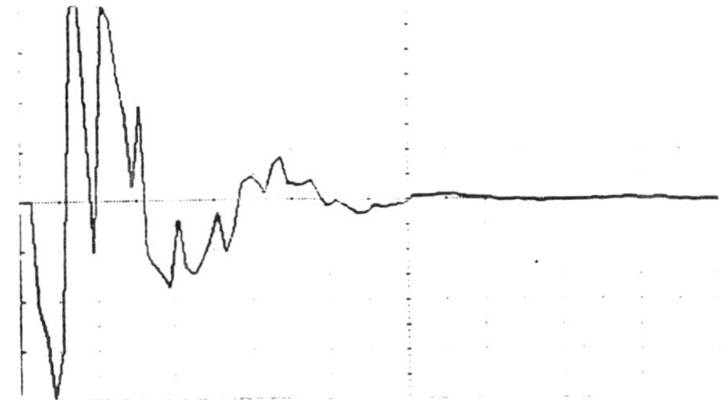

FIGURE 6: A PULSE EXCITATION WAVE FOR 1 MHZ TRANSDUCER STEEL BALL TEST

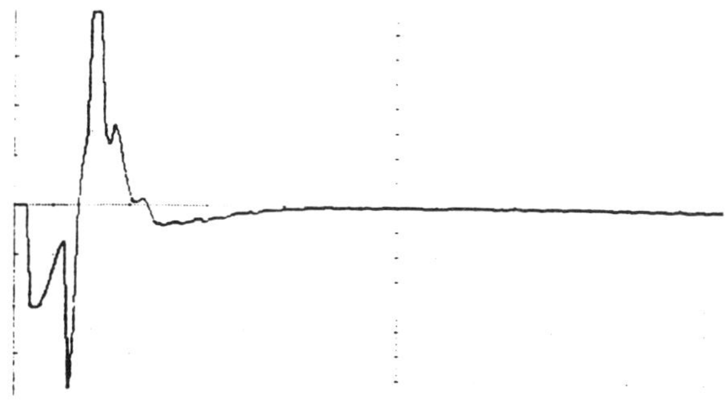

FIGURE 9: A PULSE EXCITATION WAVE FOR 5 MHZ TRANSDUCER STEEL BALL TEST

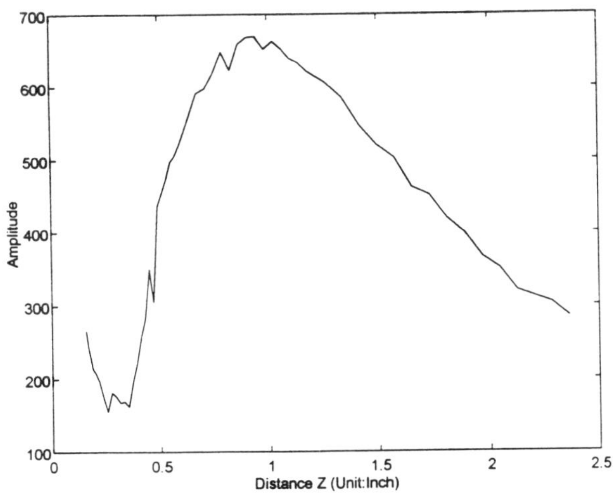

FIGURE 7: ON-AXIS SOUND PRESSURE PROFILE OF 1 MHZ FLAT TRANSDUCER

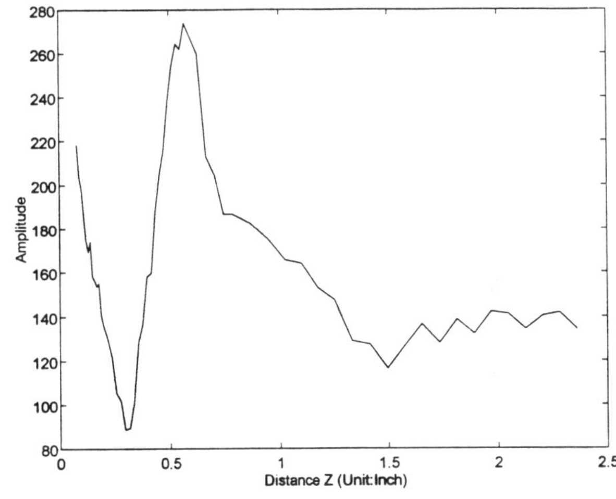

FIGURE 10: ON-AXIS SOUND PRESSURE PROFILE OF 5 MHZ FLAT TRANSDUCER

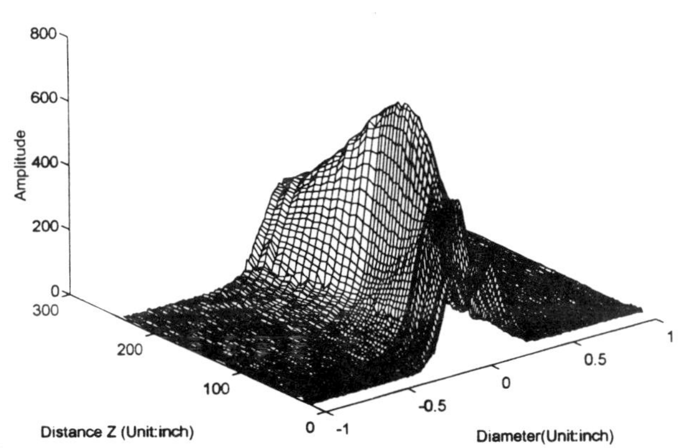

FIGURE 8: 3-DIMENSION SOUND PRESSURE DISTRIBUTION OF 1 MHZ FLAT TRANSDUCER

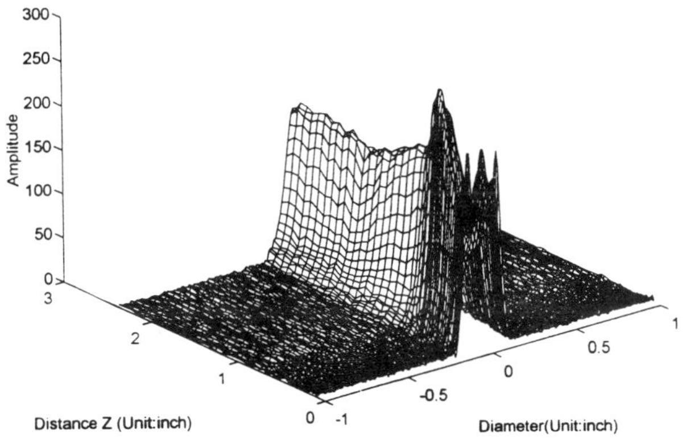

FIGURE 11 : 3-DIMENSION SOUND PRESSURE DISTRIBUTION OF 5 MHZ FLAT TRANSDUCER

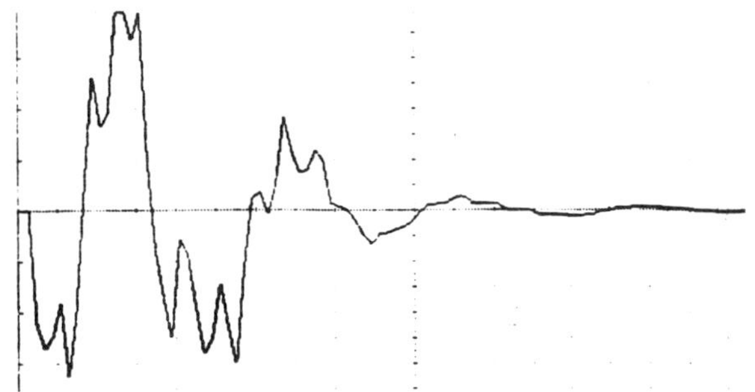

FIGURE 12: A PULSE EXCITATION WAVE FOR I MHZ
TRANSDUCER PINDUCER TEST

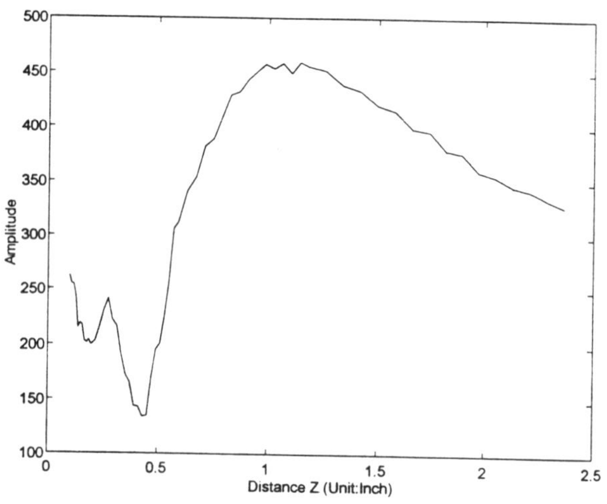

FIGURE 13: ON-AXIS SOUND PRESSURE PROFILE OF
1 MHZ TRANSDUCER

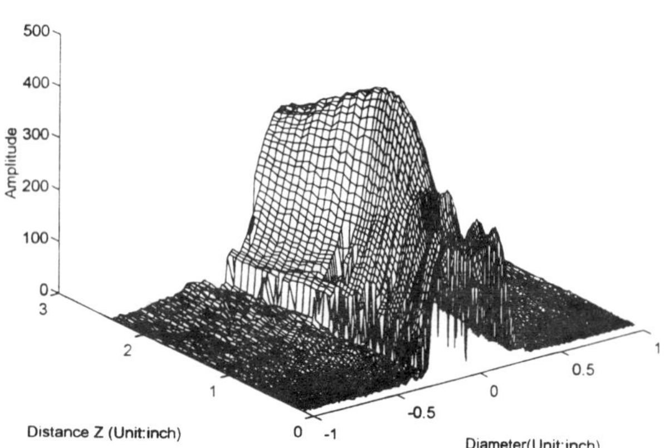

FIGURE 14 : 3-DIMENSION SOUND PRESSURE
DISTRIBUTION OF 1 MHZ FLAT TRANSDUCER

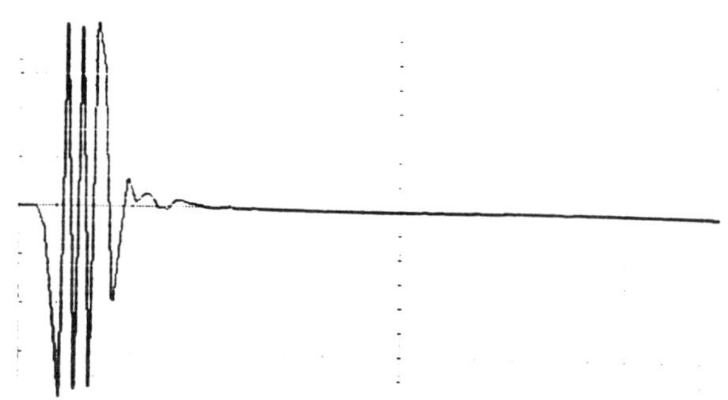

FIGURE 15: A PULSE EXCITATION WAVE FOR 5 MHZ
TRANSDUCER PINDUCER TEST

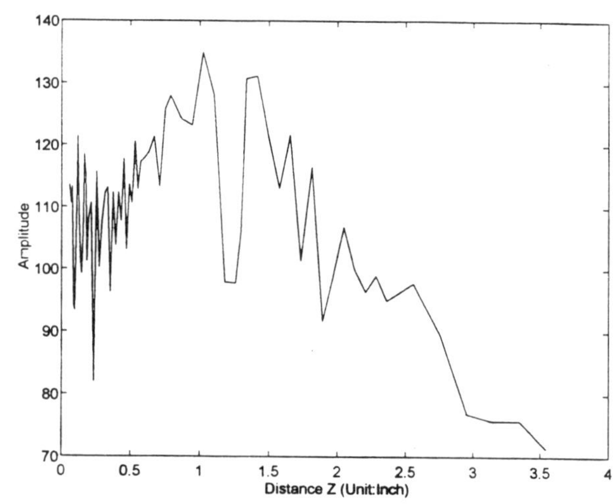

FIGURE 16: ON-AXIS SOUND PRESSURE PROFILE OF
5 MHZ TRANSDUCER

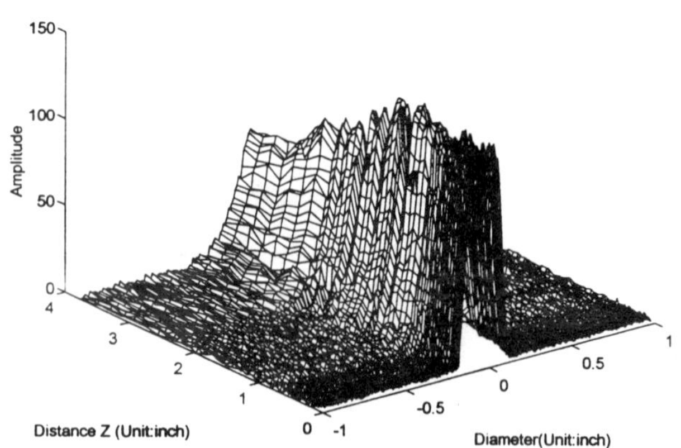

FIGURE 17: 3-DIMENSION SOUND PRESSURE
DISTRIBUTION OF 5 MHZ FLAT TRANSDUCER

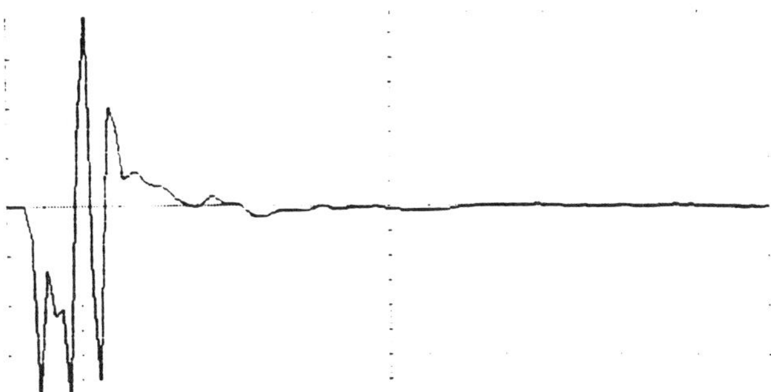

FIGURE 18: A PULSE EXCITATION WAVE FOR 5 MHZ WITH 1.46 INCH FOCAL LENGTH TRANSDUCER

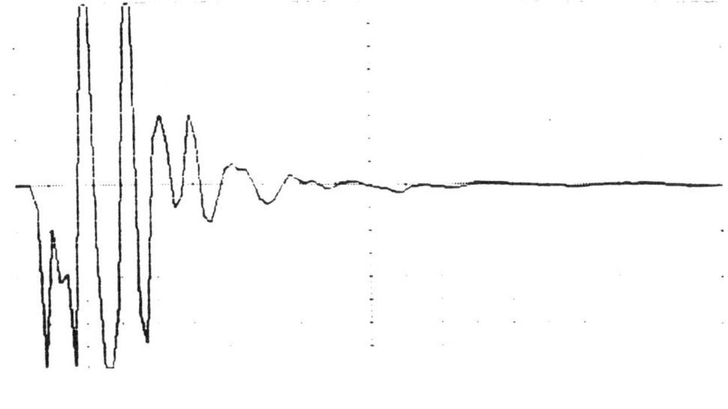

FIGURE 21: A PULSE EXCITATION WAVE FOR 5 MHZ WITH 3.15 INCH FOCAL LENGTH TRANSDUCER

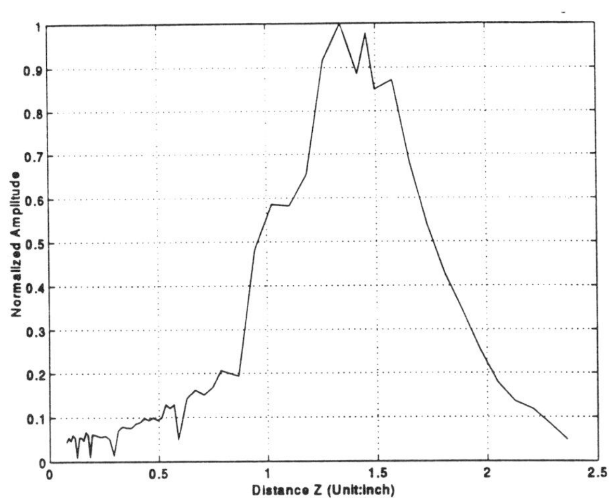

FIGURE 19: ON-AXIS SOUND PRESSURE PROFILE OF 5 MHZ WITH 1.46 INCH FOCAL LENGTH TRANSDUCER

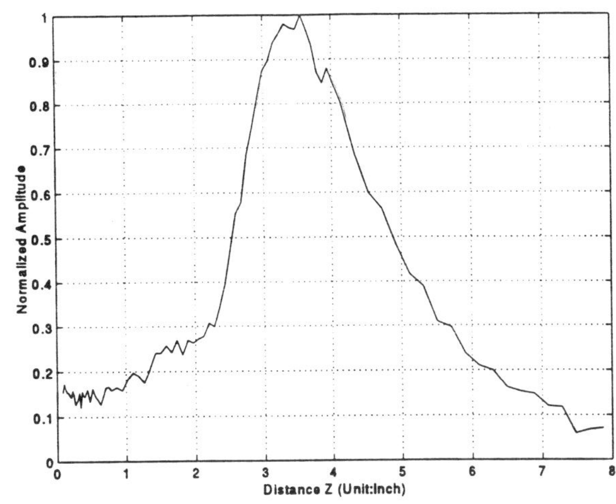

FIGURE 22: ON-AXIS SOUND PRESSURE PROFILE OF 5 MHZ WITH 3.15 INCH FOCAL LENGTH TRANSDUCER

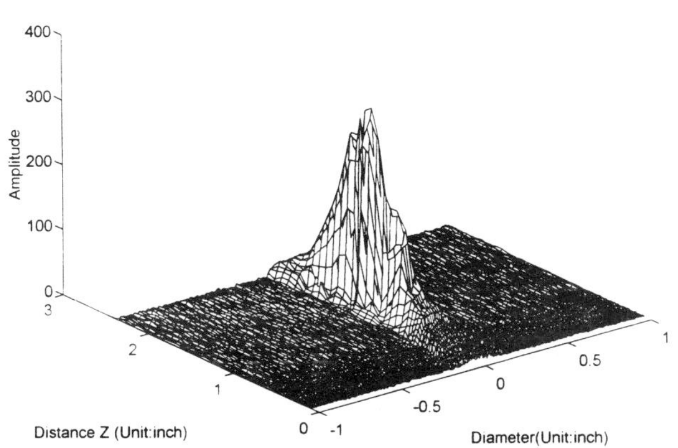

FIGURE 20 : 3-DIMENSION SOUND PRESSURE DISTRIBUTION OF 5 MHZ WITH 1.46 INCH FOCAL LENGTH TRANSDUCER

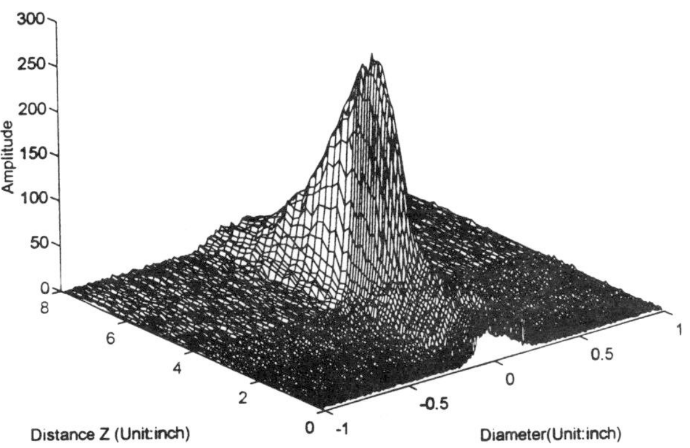

FIGURE 23: 3-DIMENSION SOUND PRESSURE DISTRIBUTION OF 5 MHZ WITH 3.15 INCH FOCAL LENGTH TRANSDUCER

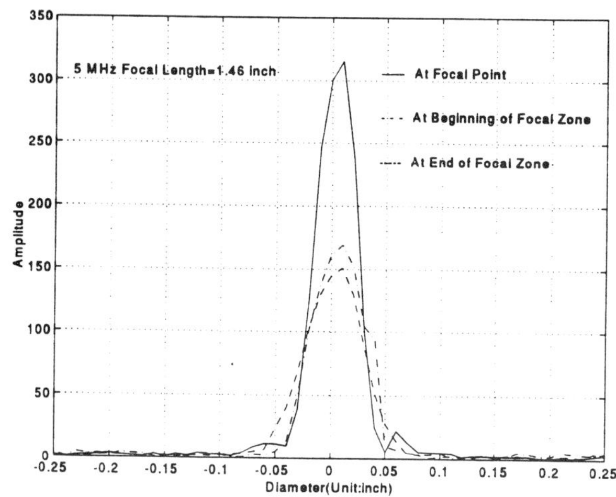

FIGURE 24: TRANSVERSE PROFILE OF 5 MHZ WITH 1.46 INCH FOCAL LENGTH FOCUSED TRANSDUCER AT THE FOCAL POINT, BEGINNING OF THE FOCAL ZONE AND END OF THE FOCAL ZONE

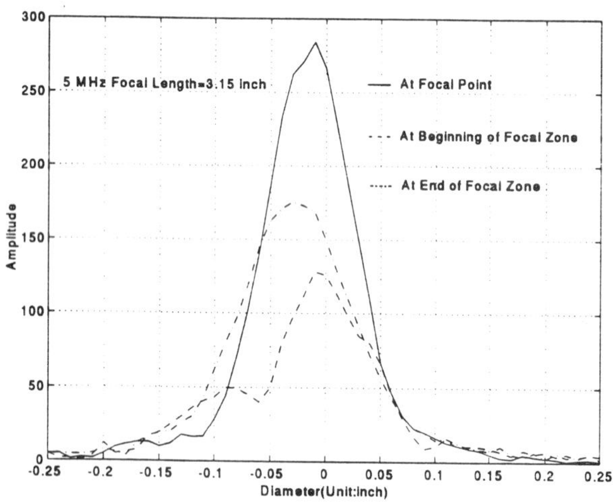

FIGURE 25: TRANSVERSE PROFILE OF 5 MHZ WITH 3.15 INCH FOCAL LENGTH FOCUSED TRANSDUCER AT THE FOCAL POINT, BEGINNING OF THE FOCAL ZONE AND END OF THE FOCAL ZONE

PRACTICAL DOMAIN FOR ULTRASONIC TESTING
OF STAINLESS STEEL OVER PLAIN CARBON STEEL
LAYERED COMPONENTS USING M_{21} WAVES

Dilawar S. Grewal and Don E. Bray
Department of Mechanical Engineering
Texas A&M University
College Station, Texas 77843-3123

ABSTRACT

The first higher order mode of the Rayleigh wave was discussed by Sezawa in the early part of this century in context of seismological wave studies. These Sezawa, or M_{21}, or first higher order mode Rayleigh waves, have subsequently been used in the field of nondestructive testing of layered materials based on the development of the seismological model of the Sezawa waves by others. In this paper the study of the Tiersten formulation in context with slow speed over high speed materials, e. g. stainless steel overlay on plain carbon steel, the limitations and applicability of that formulation is reported. This study illustrates the practical bounds for testing such layered media, using numerical analysis of this formulation for the first higher-order mode to establish theoretical limits, and corroboration of these bounds by experimental results.

INTRODUCTION

In today's industrial world an ever expanding array of materials is being used in the manufacturing of industrial products. Combining materials in the form of layers is one method used to develop components for special applications. Nuclear power generation, petroleum, railroad, chemical, composite component manufacturing, etc., industries, to name a few, utilize this technique to develop materials that are corrosion resistant, strong, or specially adapted for particularly unique applications. Inspection of layered materials with ease and reliability is a concern to these industries. Stainless steel weldment layers forming the anticorrosion lining in a simple-carbon steel three-way valve, cladding on offshore structures and pressure vessel walls, are examples of structures and components using layered media that are in need of inspection for defects.

There are a number of ways layered media can be tested for defects. Among the ultrasonic techniques available are, angled pitch-catch and normal incident ultrasonic testing methods, as addressed by the ASME Boiler and Pressure Vessel Code, Section V, T-543, 1989[1], and guided surface waves. The normal incident wave methods are extremely labor intensive, since the width of the area being tested is limited to the size of the transducer, and are also prone to missing flaws that are not oriented perpendicular to the transducer beam. The angled pitch-catch technique is slightly better, but still very labor intensive. Since the use of normal beam probes for the inspection of layered specimens is limited, due to the near field effect, angled beam probes are preferred. Angled-beam, shear probes may be used to test the material by reflecting the ultrasonic pulse off of the opposite surface. However, this method becomes ineffective due to near field effects as the probe approaches the defect. Focused probes using acoustical lenses and normal incidence probes are capable of overcoming the near field shortcomings. However, these probes cover a vary narrow area, and that translates into a very labor intensive and tedious testing protocol. Angle-beam techniques devoid of such problems include testing with surface guided waves, such as, Lamb waves, Love waves and Rayleigh waves etc. All of these techniques take advantage of the critical angle refraction of the angle-beam to send in and receive a strong ultrasonic signal.

The fundamental Rayleigh mode, M_{11}, has been used extensively to test materials for surface or near surface flaws. It will be shown in this paper that the first higher order mode, M_{21}, wave offers a more versatile and stable means of testing layered media for flaws over the M_{11} mode, as well as the method described above.

In this paper the behavior of the first higher order mode Rayleigh wave will be discussed in order to better establish a reliable protocol for the testing of layered media. ASTM A 347 austenitic stainless steel (Cr - 18%, Ni - 11% and C - 0.08%) overlay on type ASTM A387 grade structural steel (Cr - 2.25%, Mo - 1% and C - 0.25%) is used as a model for all calculations and experimentation. Practically, this model was constructed using a square block of steel measuring approximately 0.5 m (20 in.) on either side and 60 mm (2.4 in.) thick, including the cladding. The cladding was applied using a submerged arc welding process in parallel overlapping rows on 25 mm (1in.) centers. The cladding varied in thickness from 5-7 mm (0.20-0.28 in.). Dispersion patterns for the first higher order mode, M_{21}, Rayleigh waves are discussed. Subsequently experimental data is presented to corroborate the numerical results.

NUMERICAL ANALYSIS OF THE M_{21} WAVES

The M_{21} mode was first introduced by Sezawa[2] in the early part of this century. These waves were subsequently explored by various researchers in the context of seismological activity . Work by Ewing, Jardetsky and Press[3],Bolt and Butcher[4], etc. has been of great importance in establishing the basic rules for the propagation of these waves. In this paper numerical analysis of the formulation presented by Tiersten[5] is carried out to present the behavior of the M_{21} waves in a stainless steel over plain carbon steel model. In the M21 non-dispersive mode, energy is mostly conined to the top layer.

A code was written to find out the various roots for the Rayleigh wave equation using the Tiersten formulation [5]. Since the Rayleigh wave equation is a sixth order polynomial, more than one root satisfies the equation. A code was written to calculate the various roots for the equation and to separate them, in order to track individual modes of the Rayleigh wave that correspond to each one of those roots. The material properties used as input into the code for simulation of the Rayleigh modes were derived from destructive experimentation on the material. Three samples, each specimen about 16 sq. mm in cross section (4X4 mm or 0.16 X0.16 in.) and about 50 mm (1.96 in.) in length, were cut out of the block. One specimen was entirely from the base metal and the other two were taken out entirely from the cladding, one in the direction of the weldment and the other perpendicular to it. Young's modulus and bulk wave speeds in these specimens were established using the piezoelectric ultrasonic composite oscillatory technique (PUCOT). Results in the two directions showed no tangible difference. Density for the two materials was established using the Archimedes principle of water displacement. Table 1 shows the values for the above mentioned parameters used as input in the simulation of the dispersion curves for the M_{21} waves.

Table 1. Material properties for the two layers of the block.

Materials	E GPa	Poisson's Ratio	Density kg/m³
Stainless S	198	0.286	7730
Carbon S.	207	0.292	7857

The code is so designed as to consider a dependency on the values of thickness of top layer, h, incident frequency, f,

and the current value of the iterate for the phase velocity, V, for the evaluation of each step. This helps produce simulated curves for actual conditions encountered in the experimental phase. One problem encountered in solving the Bolt and Butcher formulation [4] for the evaluation of the higher order modes is that it is very difficult to simulate plots for thickness' of the top layer that are comparable to the wavelength of the incident wave, or its low multiples. The Bolt and Butcher formulation produces excellent results for earth crust models, in which the layer thickness' are huge as compared to the wavelength of the waves travelling through them. It is easier to tackle this problem with the Tiersten formulation, since it is designed for thin layers.

In both the Tiersten and the Bolt and Butcher formulation for M_{21} waves, finding the value of the first iterate for the phase velocity versus frequency curve for a given thickness involves a certain degree of speculation and experience. Each subsequent point on the plot, however, can be found by comparing to the seventh decimal place the solution of the determinant for that particular iterate and the preceding one. Larger differences between two subsequent iterates may result in confusion between various roots.

Figure 1 shows the simulated curves derived from the Tiersten formulation for the variable values shown in Table 1.

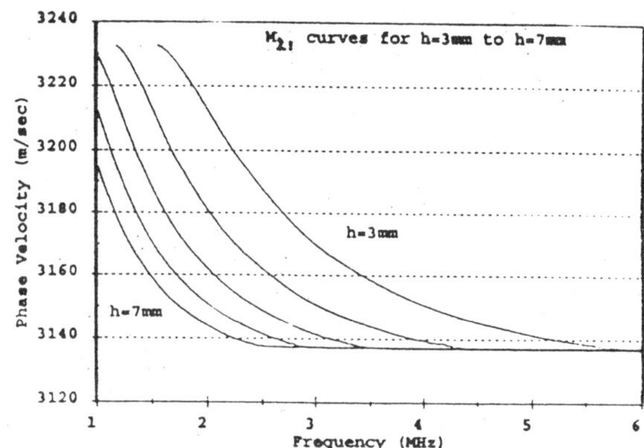

Figure 1. Phase velocities for M_{21} waves for a stainless steel overlay on plain carbon steel model.

The following observations can be made from the plot in Figure 1:
1). The M_{21} waves are not dispersive for higher values of the excitation frequencies for a given top layer thickness. Hence, in a practical situation, it would be helpful to know in advance the frequencies for which these waves are dispersive or non-dispersive.
2). It can be noted that at higher frequencies the velocity of the M_{21} waves approaches that of the bulk shear wave in the top layer.
3). For any given top layer thickness (h), there exists a cutoff frequency below which the M_{21} wave does not exist. The values of observed cutoff frequencies for various top layer thickness' for this particular model are shown in Table 2. Hence, plotting the simulated curves before choosing the testing fre-

quency for a given setup would ensure that optimum results are achieved without involving much guess work.

Table 2. Cutoff frequencies for simulated M_{21} waves.

Thickness (mm)	3	4	5	6	7
Cutoff Frequency (MHz)	1.54	1.17	0.93	0.78	0.72

EXPERIMENTAL RESULTS

Earlier studies by Hausler[6] and Bray, Egle and Reiter[7] showed that the M_{21} waves can easily be generated using a commercially available surface wave clear acrylic wedge. A variable angle wedge was used first in this study to evaluate the most optimum angle for the generation of M_{21} waves in the stainless steel layer. Transducers with nominal center frequencies of 0.5, 1, 2.25, 3.5 and 5 MHz were used in conjunction with the variable angle wedge to generate the M_{21} wave in the specimen. For each incident frequency travel-time measurements were taken for various lengths by choosing the same point at the beginning of the returned M_{21} wave signal, or as close to it as possible. Travel time in the plexiglas wedge was eliminated by subtracting one travel-time measurement from another and then calculating the resultant speed. A number of readings were taken for a representative average value, so that chance errors may be avoided. Observed travel times are noted in Table 3.

Table 3. Experimental travel-time data for M_{21} waves.

Dist. \ Freq. cm (in.)	1 MHz.	2.25 MHz	3.5 MHz	5 MHz
12.7 (5)	70.179 µs		72.304 µs	71.04µs
20 (7.87)		141.825 µs		
25.4 (10)	111.79 µs		114.695 µs	112.05µs
30 (11.8)		176.605 µs		
AVE. VEL.	3052 m/s	2875 m/s	2996 m/s	3097 m/s

In order to calculate the group velocities from phase velocities depicted in Figure 1, first the slope of the curve is calculated for every two consecutive iterates, and then the group velocity derived. The M_{21} wave velocities from Table 3 are plotted along with the theoretical simulated group velocities in the plots shown in Figure 2. The theoretical curves have been extended manually since it becomes extremely difficult to measure the slope of each curve for a given iterate when slight fluctuations occur in the data, resulting in a lot of "noise". The overall curve still remains similar to the one where no noise has been eliminated. Even though the thickness of the stainless steel layer is not uniform in the specimen, it can be seen from Figure 3 that the experimental points coincide well with the theoretical plots.

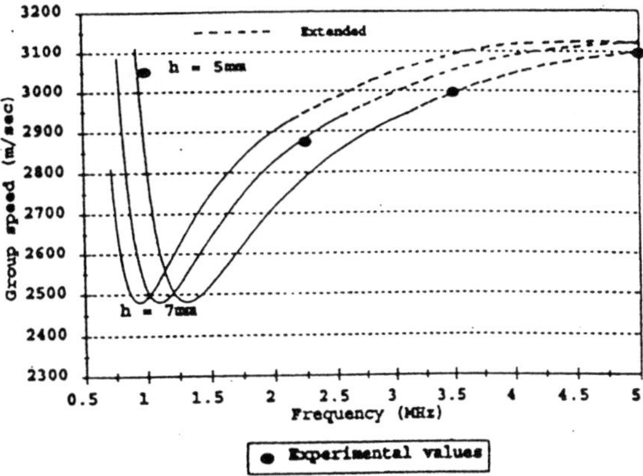

Figure 2. Observed experimental M_{21} wave speeds and Group velocities for M_{21} wave as derived from the phase velocities illustrated in Figure 1.

In addition to the determination of wave speeds for the M_{21} wave, the experimentation also revealed that for incident angles of 60^0 to 64^0, both the M_{11} and the M_{21} can be generated. The M_{21} is strongest at a generation angle of about 62^0. For 64^0 to 74^0, the M_{11} is strongest and the M_{21} loses most of its energy. Hence, depending upon the application, a slight variation in the incident angle can be used to obtain the resulting wave or waves by controlling the amount of energy being transferred into that wave.

Figures 3 and Figure 4 show the actual signals received while exciting the surface waves in the aforementioned specimen. An important distinction between the M_{11} and M_{21} waves can be established by observing the signals in the two figures. Even though the peak amplitude of the M_{11} is greater than that of the M_{21} in Figure 3, if the surface of the specimen is touched, the M_{11} is damped much more drastically than the M_{11} signal. This is helpful in that there is no need for surface preparation of the specimen for supporting the M_{21} waves. Taking the example of the current model, the surface has to be ground for the M_{11} to be able to exist in the specimen. However, there is no need to grind or polish the surface if the specimen is to be tested using the M_{21} waves. A rough surface actually aids in the testing procedure, since only one signal is returned. Care, however, has to be taken in providing enough contact area between the probe and the specimen, so that a tangible amount of energy can be put into the top layer of the specimen. This attribute of the M_{21} waves can be of great utility when testing either welded media or specimens that have been shot-peened etc. A lot of time and money can be saved by reduced surface preparation requirements. Hirao, Fukuoka and Toda[8] have utilized the dispersion patterns of surface waves to non-destructively evaluate the hardening depth in surface treated steels.

SUMMARY

M_{21} wave is a very versatile tool that can be utilized for not only testing layered media for flaws, but also to evaluate the elastic properties of a material. Using the theoretical curves for a given set of materials, the practical domain for testing the specimen can be easily established.

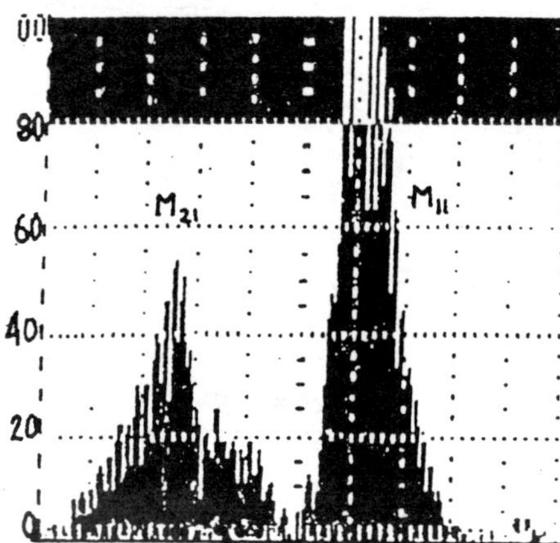

M_{21} arrival at 111.79μs for a distance equal to 25.4 cm (10 in.)

Figure 3. Undamped surface wave signal in stainless steel overlay on carbon steel specimen.

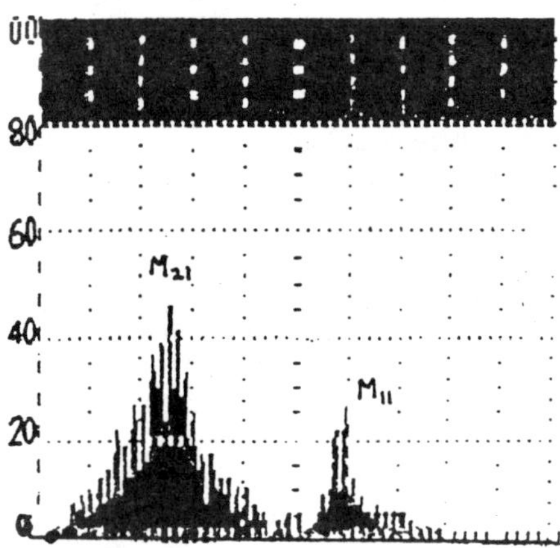

Figure 4. Damped surface wave signal in stainless steel overlay on carbon steel specimen.

Conversely, if experimental data using narrow band transducers is available, a rough estimate of the physical properties of the top layer material possibly may be achieved by comparison with theoretical curves. Also, these waves can be used to ascertain the depth and condition of surface hardened layers, claddings and cold-worked layers etc. Since the M_{21} is a mode of the Rayleigh wave, they inherit the characteristics of Rayleigh waves, i. e. these waves can follow contours, just as the fundamental mode waves do. Using the M_{21} waves a lengths upto 10 inches or more can be tested in one pass, saving labor and reducing the number of decisions the inspector has to make. This is one major advantage over normal incidence techniques. Also, surface preparation is of lesser consequence as compared to other ultrasonic testing techniques.

ACKNOWLEDGEMENTS

The authors would like to acknowledge and thank CBI, Houston, Texas, for manufacturing and furnishing the stainless steel overlay plate used in this study.

REFERENCES

1. The American Society of Mechanical Engineers, ASME Boiler and Pressure Vessel Code (New York, New York, 1989), Section V, Nondestructive Examination, Section V Topic T-543.

2. Sezawa, K. (1927). "Dispersion of Elastic Waves propagating on the Surface of Stratified Bodies and on Curved Surfaces," Bulletin Earthquake Research Institute, Tokyo University, 3, 1-18.

3. Ewing, M., Jardetsky, W., Press, F., Elastic waves in layered media (McGraw-Hill, New York, 1957).

4. Bolt, B. A., and Butcher, J. C. (September 1960), "Rayleigh Wave Dispersion for a Single Layer on an Elastic Half Space." Australian Journal of Physics, 13, 3, 498-504.

5. Tiersten, H. F., (February 1969), "Elastic Surface Waves Guided by Thin Films." Journal of Applied Physics, 13, 770-780.

6. Hausler, E., (1960), "Elastic Waves on the Surface of a Solid, Especially Stratified Media, and Their Application to Nondestructive Testing," Materialpruf, 2, 51-55.

7. Bray, D. E., Egle, D. M., and Reiter, L., (1978), "Rayleigh Wave Dispersion in the Cold-worked Layer of Used Railroad Rail," Journal of the Acoustical Society of America, 64, 3, 845-851.

8. Hirao, M., Fukuoka, H., Toda H., Sotani, Y., and Suzuki, S., (1983), "Non-destructive Evaluation of Hardening Depth Using Surface-Wave Dispersion Patterns," Journal of Mechanical Working Technology, 8, 171-179.

NDE OF HEAVY WALL STAINLESS STEEL CRYOGENIC ASME PRESSURE VESSEL WELDS

C. N. Sherlock, R. W. Kruzic, H. K. Howerton, and J. R. Phillips*
Chicago Bridge and Iron Company
Houston, Texas

ABSTRACT

This article is a discussion of how changes in the orientation of shell welds in a heavy wall stainless steel pressure vessel impacted the radiography and ultrasonic methods used to volumetrically examine these welds. Particular emphasis is on the fabrication of the ultrasonic calibration block(s) which were made with a replication of the welds to be examined to obtain valid and reliable UT results for this type material.

This article also discusses the selection and performance of a video system used for the in-process welding control of the deep narrow gap welds in the 15 1/2" (394mm) thick stainless steel inner vessel shell.

BACKGROUND

In late 1989, a contract was awarded to CBI to manufacture a double wall vacuum insulated high pressure liquid hydrogen (LH_2) vessel. The inner spherical vessel and the outer cylindrical vessel were to meet the respective requirements of the ASME Section VIII Division 2 and ASME Section VIII Division 1 Boiler and Pressure Vessel Code. The inner vessel was designed with an inside diameter of 130.2 inches (331cm) and a capacity of 5000 gallons (19 m^3) of LH_2. It was designed for a pressure of 8500 psig (58,600 kPa) with a full external vacuum over a temperature range of minus 423°F (-253°C) to plus 140°F (60°C) [37°R (20 K) to 600°R (333 K) absolute]. The cylindrical outer vessel was designed with hemispherical heads having an inside diameter of 221.25 inches (562cm) with a 42 inch (107cm) long thick straight shell between the heads. It was designed to withstand a full internal vacuum over an ambient temperature range. See Figure 1.

The design pressure and temperature for the spherical inner vessel required that it be fabricated of 15 1/2" (394mm) thick SA 336 Gr F347 stainless steel (SS) forged material. The cylindrical outer vacuum vessel was designed and fabricated using SA 516 Gr 60 carbon steel plate.

The manufacturer originally planned to fabricate the heavy wall stainless steel inner vessel from two hemispherical forgings. This would have resulted in one complete circumferential equator weld whose axis would have been oriented on a radial centerline of the vessel. To fulfill the requirements of the ASME Section VIII Division 2 Code, the radiography of this weld would have been routine with the beam of radiation normal to the weld axis. The customer specified ultrasonic angle beam examination of this radial weld would have been fairly difficult to perform due to the thickness and the narrow gap weld configuration with the weld sidewall fusion lines being nearly perpendicular to the forging surface.

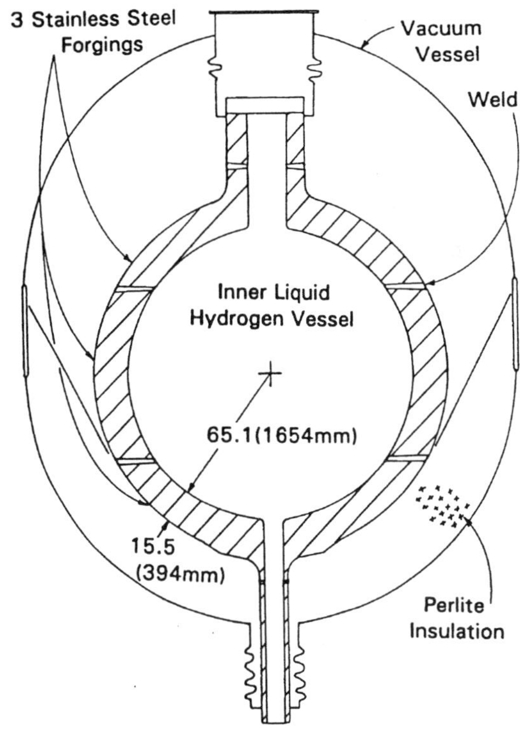

Figure 1 DOUBLE WALL HIGH PRESSURE 5000 GALLON LH_2 VESSEL

The size of the stainless steel hemispherical forgings exceeded the capacity of forging suppliers. This necessitated a re-design of the inner vessel using three forgings to be joined with two weld joints whose axes were not on a radial centerline of the vessel. These two weld joints were detailed for a narrow gap submerged-arc welding (SAW) process[1]. To have a weld joint with a vertical orientation during welding, each of the two welds was oriented with its axis at an angle with the radial lines to the center of the vessel. This resulted in both welds being approximately 18.75" (476mm) thick at their centerline as seen in Figure 2.

NDE OF THE INNER VESSEL SHELL AND PENETRATION WELDS

Visual (VT)

The deep narrow gap shell weld joint configuration made it very difficult for the welding operator to directly view the weld puddle during welding. A commercial off-the-shelf video system with bright flood lighting enabled the operator to view the in-process weld surface remotely. This video system provided the continuous remote visual inspection of the in-process welding that was necessary to verify sidewall fusion. Lack of sidewall fusion is evidenced by undercutting which could be detected by the welding operator monitoring the video display.

While viewing the in-process welding of a 15.5" (394mm) thick test plate to produce the Procedure Qualification Record (PQR) and the ultrasonic calibration blocks for the inner vessel shell, it was determined that interpass liquid penetrant examinations would not be necessary.

After each weld pass, the cleaned weld surface was again visually inspected for discontinuities; in particular any signs of undercutting that could indicate lack of sidewall fusion.

Radiography (RT)

The two circumferential non-radial weld shell joints were radiographed as shown in the exposure set-up in Figure 2 by taking overlapping radial exposures in 7" to 8" (178 to 203mm) intervals using fine grain (what used to be called Class 2) film with 20 (sometimes 30) mil front screens and 20 (sometimes 30) mil back screens. Because of the angle of the weld to the beam of radiation and the film, the film had to be special ordered to a width of 22" (560mm) to ensure complete coverage across the projected width of the weld. The high energy x-ray radiation source was an 8 Mev linear accelerator with a 0.079" (2mm) focal spot. With the unit operating at full capacity at a source to film distance of 72" (1829mm), the exposure time for a density of 2.5 to 3.0 was four (4) hours. Unsharpness for the 15.5" (394mm) thick material was 0.022" (0.55mm).

The types of discontinuities that would most easily be revealed with radial exposures of these non-radial weld joints would be voids caused by slag or rounded or elongated porosity indications. Discontinuities such as lack of sidewall fusion or longitudinal cracks would be very difficult to detect in radiographs produced with the beam of radiation not normal to the weld axis.

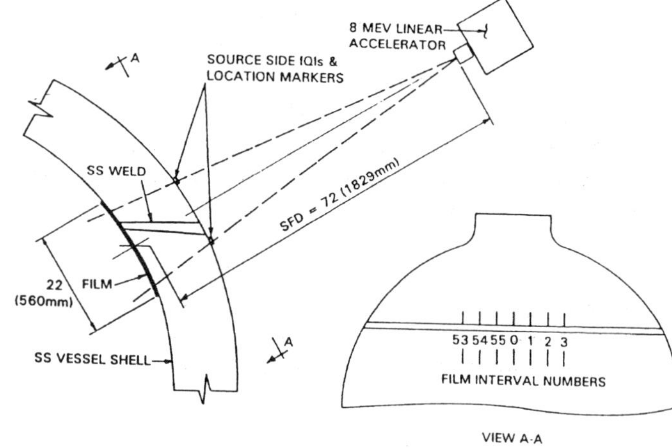

Figure 2 RADIOGRAPHIC EXPOSURE SET-UP

These exposures made with the beam of radiation not normal to the weld axis were considered acceptable on the basis that:

- this weld material and weld process tend to produce high quality welds free of sidewall fusion or cracking when closely controlled.
- lack of sidewall fusion, cracks, slag inclusions or other discontinuities would have been detected during the continuous remote visual inspection of the in-process welding.
- lack of sidewall fusion and cracks would be detected during the supplementary volumetric ultrasonic angled scans transverse (perpendicular) to the weld axis.

Ultrasonics (UT)

Because of the nature of the radiography exposures of these vessel shell welds, an ultrasonic examination of these welds would have been performed on these non-radial welds by this manufacturer regardless of the requirements of the customer specification.

Due to the grain structure of austenitic welds, the following factors are particularly important when examining these welds:

- Angle of incidence of sound beam relative to the weld interface.
- Distance of the sound path, particularly in the weld metal.
- Frequency of the sound beam. In this 15.5" thick austenitic weld metal, a 1.5 MHz frequency produced the best results for this manufacturer. For thinner penetration welds, 2.25 Mhz produced the best results.
- Mode of vibration of the sound wave. Focused longitudinal wave normal incidence and angle incidence dual element search units were used. This is the only ultrasonic technique that has proven effective for this manufacturer when examining SS austenitic welds.
- Multiple focal length (roof angle) search units for these thicker materials to achieve an acceptable examination of the total thickness.

Because of the shell curvature, the design change from one radial to two non-radial circumferential stainless steel shell welds resulted in an average incidence angle more perpendicular to the weld centerline for the selected 45° angled transverse (perpendicular) scans. For the most part, this shortened the sound path through both the forging and weld metal which reduced the attenuation for these scans. Conversely, it lengthened the sound path through both the forging and the weld metal for the selected 45° angled scans in the direction of the weld. However, it was anticipated that the attenuation for the scans in the direction parallel to the weld axis would be excessively high, preventing a successful ultrasonic examination in this direction. That later proved to be the case.

UT Calibration Blocks

For 15.5" (394mm) thick material, the ASME Section V Code, Article 5 requires 9/16" (14.3mm) ∅ side drilled holes (SDHs) in the UT calibration block. This Code also requires that the basic calibration block material be of the same product form or equivalent P number grouping. Code paragraph T-542.8.5, states that it may be necessary to modify the calibration block for austenitic alloy welds. From previous ultrasonic experience with the large unpredictable attenuation and associated sound beam effects caused by the grain structure of austenitic weld metal, this manufacturer knew that it was necessary that the calibration block contain a duplicate of the vessel welds with the SDHs in the weld metal in order to prove

penetration of the sound beam into those welds. To increase the sensitivity level and provide more calibration reflectors, a plan was developed to treat the 15.5" (394mm) thick inner vessel shell weld ultrasonic calibration block as though it consisted of six zones each equal to a T of about 2.6" (66mm) instead of a weld thickness T of 15.5" (394mm). This same Code requires 3/16" (4.8mm) ∅ SDHs for a weld thickness over 2" (51mm) through 4" (102mm). Thus, shell basic ultrasonic calibration blocks for the transverse (perpendicular) weld scan calibration were made by drilling 3/16" (4.8mm) ∅ 1/4T, 1/2T and 3/4T holes into the weld in the direction of the weld axis in each of the six zones for a total of 18 holes. See Figures 3 and 4.

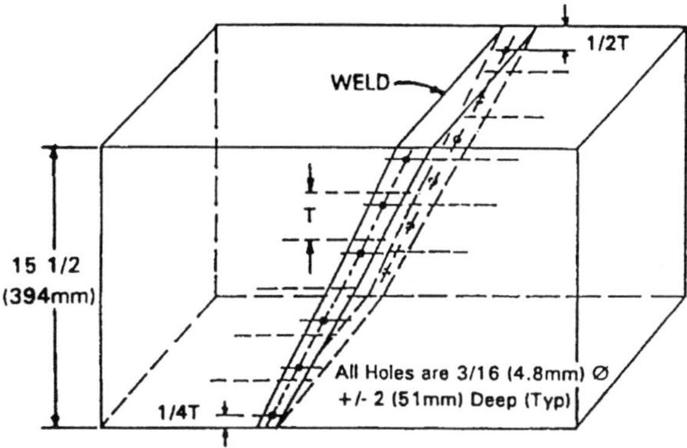

Figure 3 UT CALIBRATION BLOCK WITH ZONED 1/4T AND 1/2T HOLES DRILLED IN THE DIRECTION OF THE WELD

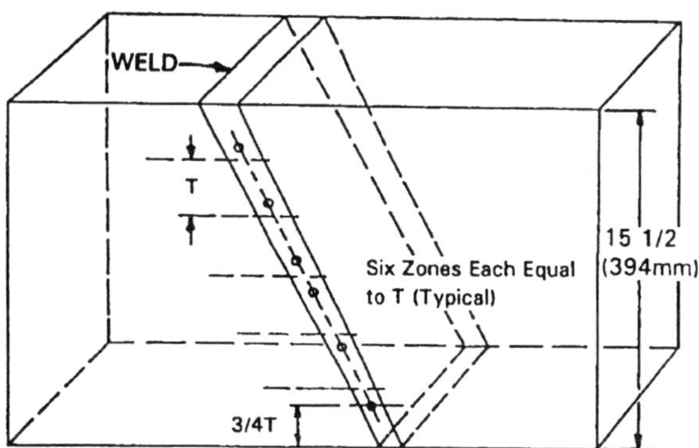

Figure 4 UT CALIBRATION BLOCK WITH ZONED 3/4T HOLES DRILLED IN THE DIRECTION OF THE WELD

A third block for the longitudinal (parallel) weld scan calibration was made by drilling 3/16" (4.8mm) ∅ 1/4T, 1/2T and 3/4T into the weld in a direction transverse (perpendicular) to the weld axis. See Figure 5.

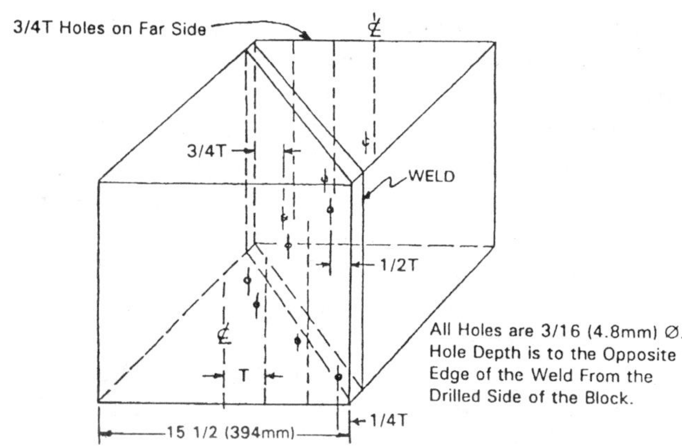

Figure 5 UT CALIBRATION BLOCK WITH ZONED 1/4T, 1/2T AND 3/4T HOLES DRILLED TRANSVERSE TO THE WELD

This same plan was used to make the 9" (229mm) thick UT calibration blocks for the 18" (457mm) I.D. manway. See Figure 6.

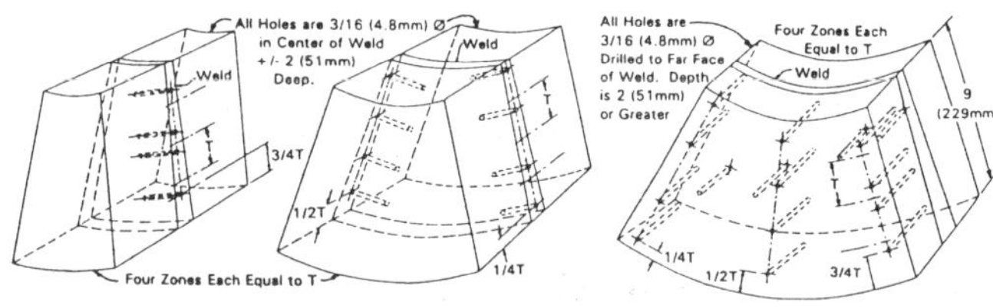

Figure 6 MANWAY UT CURVED CALIBRATION BLOCKS

For smaller diameter thinner wall penetrations, the normal ASME Code SDH requirements were used. An example is shown in Figure 7.

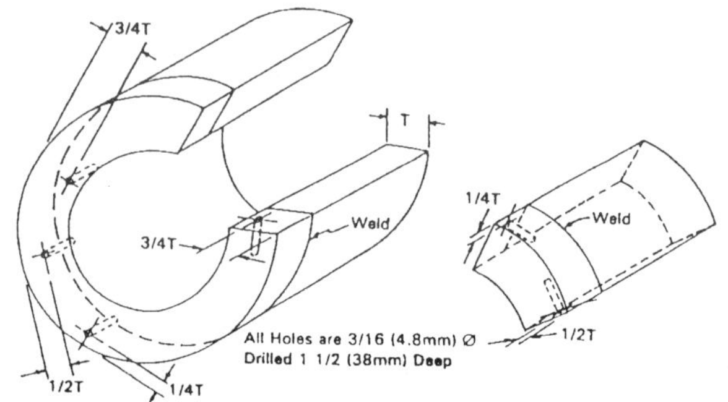

Figure 7 TYPICAL NOZZLE UT CALIBRATION BLOCKS

Photograph #1 shows the UT stainless steel ASME calibration shell weld blocks and penetration weld blocks that were developed for this vessel.

Photograph 1 UT CALIBRATION BLOCKS FOR INNER VESSEL

Normal Incidence Focused Scans

For the normal incidence shell weld scans, it was only necessary to use a short focus wedge (roof angle of 198°) and a long focus wedge (roof angle of 186°) to obtain complete depth coverage of the entire area adjacent to both sides of the weld. Photograph #2 shows the normal incidence long focus search unit set up on the calibration block over the 3/4T hole in the inside zone.

Photograph 2 NORMAL INCIDENCE SEARCH UNIT ON SHELL BLOCK

Because the welds were at an angle to the shell surface, these normal incidence scans had to cover a wide area transverse (perpendicular) to the weld axis in order to extend beyond the sides of the weld on both surfaces. As shown in Figure 8, two scans were made from the outside shell surface to examine the parent metal in the area beyond one side of the weld and two scans were made from the inside shell surface to examine the parent metal in the area beyond the other side of the weld. On each surface one scan was made with the crosstalk barrier of the dual transducer search unit in a direction perpendicular to the weld axis and the second scan was made with the crosstalk barrier in a direction parallel with the weld axis. In Figure 8 the required two direction normal incidence scans are shown as examination "A_1 and A_2" from the outside surface and as examination A_3 and A_4 from the inside surface.

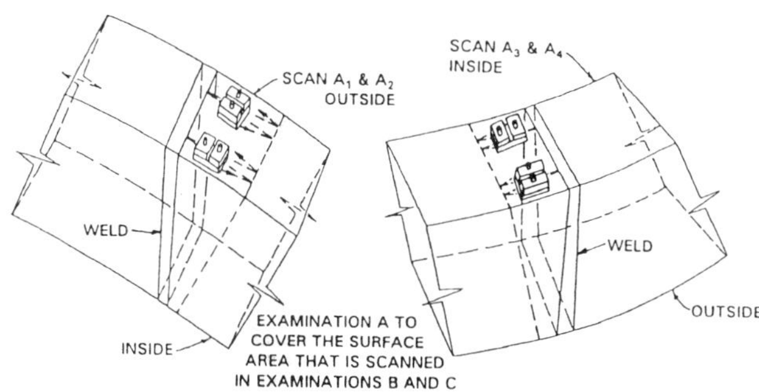

Figure 8 NORMAL INCIDENCE SCAN PATTERNS

Angled Focused Scans

Because of the variation in the angle from the shell surface to the weld interface due to the curvature of the shell and the manufacturer's knowledge of sound beam effects versus incidence angle to the coarse grain weld metal, a 45° focused longitudinal sound beam was selected. This was confirmed during the procedure demonstration to be the best overall choice for a single angle transverse (perpendicular) scan.

The procedure demonstration for the transverse (perpendicular) scan showed that complete coverage to a depth of 8" (203mm) could be obtained with two (2) search units; one short focal length and one long focal length. Photograph #3 shows the short focal length search unit set up over the 3/4T hole in the outside zone. Electronic linearization of the distance amplitude curve (DAC) at 80 % of full screen height (FSH) for each focus unit over a depth of more than 4" (102mm) was readily achieved. Based on this, it was decided that the best approach for this portion of the examination was to perform one shell surface scan from the short sound path side of the weld axis for a depth of T/2 + 1/2" = 8" (203mm) and then do a second opposite shell surface scan from the short sound path side of the weld to a depth of 8" (203mm). This provided an

area of overlap to ensure a complete transverse angle beam examination. These are shown as scans B_1 and B_2 in Figure 9.

Photograph 3 ANGLED SEARCH UNIT ON SHELL CALIBRATION BLOCK FOR TRANSVERSE SCAN

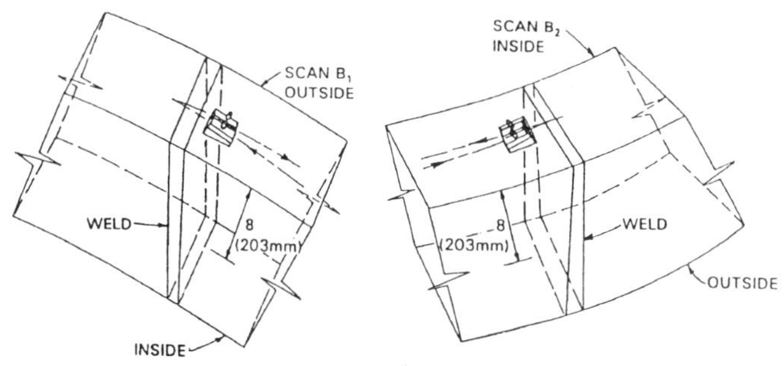

Figure 9 ANGLED TRANSVERSE (PERPENDICULAR) SCAN PATTERNS

The procedure demonstration for the longitudinal scan showed very large attenuation. The cathode ray tube (CRT) amplitude from the first hole was visible, but the amplitude from the second hole was barely distinguishable above the grass. Because of this result and the lack of probability of the existence of transverse indications with this weld configuration and welding process, it was decided to omit the longitudinal (parallel) angled focused scan from the procedure. This portion of the ultrasonic examination was later added to the procedure at the customer's request and performed even though the manufacturer's personnel felt it was not accomplishing anything significant ultrasonically. These are shown as scans C_1 and C_2 for one weld interface in two directions and scans C_3 and C_4 for one weld interface in two directions both to a depth of 8" (203mm) in Figure 10.

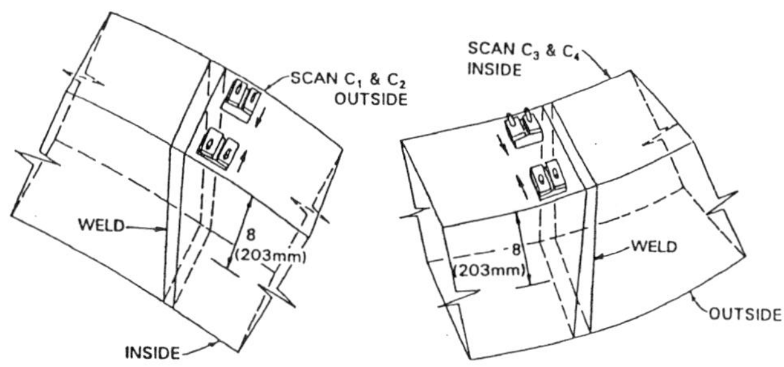

Figure 10 ANGLED LONGITUDINAL (PARALLEL) SCAN PATTERNS

RESULTS

Remote Visual

No undercut, indicating lack of fusion, was seen during the continuous remote visual inspection performed during the welding operation of both inner vessel shell SS welds.

Radiography

No rejectable indication was revealed in any of the radiographs.

Ultrasonics

During the calibration on the shell UT stainless steel calibration blocks there was a grain boundary indication caused by one area in the weld where there were several weld passes just slightly larger than the surrounding normal size weld passes. The spot in the block weld from which the CRT indication was originating was marked for location on the side of the block. That area was then etched and the grain boundaries in that spot clearly showed the extra large weld passes.

In the entire length of both welds, no CRT signal was indicated that had to be recorded (above 50% of the DAC) or had to be investigated (above 20% of the DAC) during the ultrasonic examination of both welds.

Helium Mass Spectrometer (MSLD) Test

Both the inner and outer vessels were helium mass spectrometer tested with no detectable leakage indicated. The outer vessel was evacuated and HMS tracer probe leak tested followed by a pressure rise (vacuum retention) test. The inner vessel was then pressurized with a helium mixture and quantitatively HMS leakage rate tested with HMS sampling the evacuated annular space between the vessel.

Structural Integrity (SI) Test

Both the inner and outer vessels were successfully overload tested in accordance with the respective requirements of the ASME Section VIII Division 2 and ASME Section VIII Division 1 Boiler and Pressure Vessel Code.

1 "Narrow-Gap Subarc Builds Thick-Walled Tank" by Roger Daemen and Ron Shockley, published in Welding Design & Fabrication, May 1992, Pages 30 and 31.

* During the time this vessel was being constructed, authors Sherlock and Kruzic worked in Chicago Bridge & Iron Company's NDE group of Corporate Welding, Houston, Texas. Howerton and Phillips worked in the CBI Services Company's Welding & NDE department at the Manufacturing facility in Cordova, Alabama where this vessel was built. Mr. Phillips presently works for Law Engineering in Birmingham, Alabama.

NDE-Vol. 13, NDE for the Energy Industry
ASME 1995

STRESS MEASUREMENTS IN THE FIELD ON
A NATURAL GAS PIPELINE VIA X–RAY DIFFRACTION

James A. Pineault and Michael E. Brauss
Proto Manufacturing Limited
Oldcastle, Ontario, Canada

ABSTRACT

X-ray diffraction techniques can be applied in the accurate characterization of residual and applied stresses. In this case study, x-ray techniques were used to measure the stresses present in a natural gas pipeline under both pressurized and depressurized conditions. The technique measures the absolute or real stress present in the system in both cases. Hence, stresses due to working pressures, fabrication, shot blasting or peening and those stresses due to the pipe bending under its own weight are all measured. Furthermore, stresses in the pipeline due to environmental effects such as earth movement and erosion are also measured.

The accurate characterization of residual and applied stresses in natural gas pipelines is essential for a successful preventative maintenance program. The importance of detecting conditions where distortions exist and cracks can easily initiate and propagate, particularly those conditions that may ultimately lead to catastrophic failures, cannot be underestimated.

For field measurements of both residual and applied stresses, there is no other nondestructive method known that is as accurate, versatile or practical than the x-ray diffraction method.

This paper will describe the experiments and the results of stress measurements performed on a natural gas pipeline.

INTRODUCTION

The x-ray method does not measure stress directly but measures strain from which stress values are calculated. The x-ray method rather elegantly takes advantage of the crystalline structure of the material itself, by using the atomic lattice spacing as a strain gage. As a result, thousands of "built in strain gages" within the crystals which compose the material are available for strain measurement by the x-ray diffraction method. To phrase it more exactly, the surface strain present can be determined by the measurement of the elastic atomic lattice spacing or "d-spacing" as it is commonly called. This lattice spacing, the distance between the planes of atoms, is a function of the material and the stresses present in the material. The x-ray diffraction angle θ for a given x-ray wavelength λ can be used to determine the material "d" spacing by solving Bragg's law:

$$n\lambda = 2d\sin\theta \dots\dots\dots\dots\dots\dots\dots\dots\dots\dots\dots\dots\dots\dots\dots(1)$$

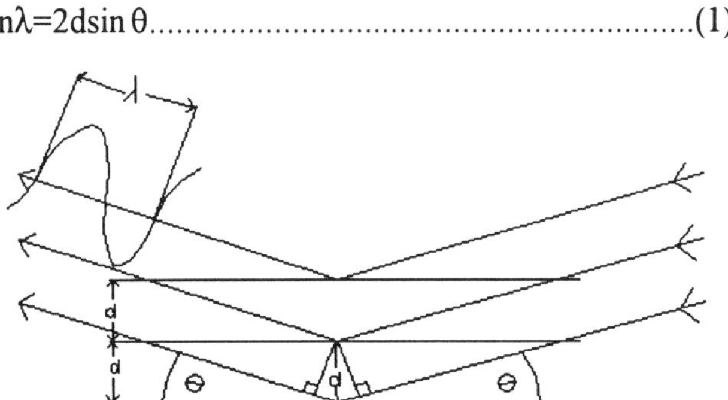

FIGURE 1 - WHEN THE PATH DIFFERENCE OF TWO MONOCHROMATIC INCIDENT WAVES IS AN INTEGER MULTIPLE OF THE RADIATION WAVELENGTH λ, CONSTRUCTIVE INTERFERENCE MAY OCCUR IN THE CRYSTAL.

For x-ray diffraction to occur, i.e. constructive wave interference, the path difference traveled by the diffracted beam through the material, as compared to a non-diffracted beam, must be equal to $n\lambda$ (Noyan and Cohen, 1987). The presence of residual stresses in the material produces a shift in the x-ray diffraction peak angular position (Cullity, 1978) which is directly measured by the detector.

Once the lattice d-spacings are measured for the unstressed (d_o) and stressed (d_i) material conditions, the atomic lattice strain can then be calculated by the following relationship (Hilley et. al., 1971):

$$\text{strain} = (d_i - d_o)/d_o \quad\dots\dots\dots\dots\dots(2)$$

For isotropic materials, strains can be converted to stress values using the equation shown below.

$$\text{stress }(\sigma) = \frac{d_\psi - d_O}{d_O}\left(\frac{E}{1+\upsilon}\right)\frac{1}{\sin^2\psi} \quad\dots\dots\dots\dots(3)$$

where: $\frac{(E)}{(1+\upsilon)}$ is the x-ray elastic constant, ψ is the angle subtended by the bisector of the incident and diffracted beam and the surface normal, d_ψ is the lattice spacing at a given ψ tilt and d_o is the unstressed lattice spacing.

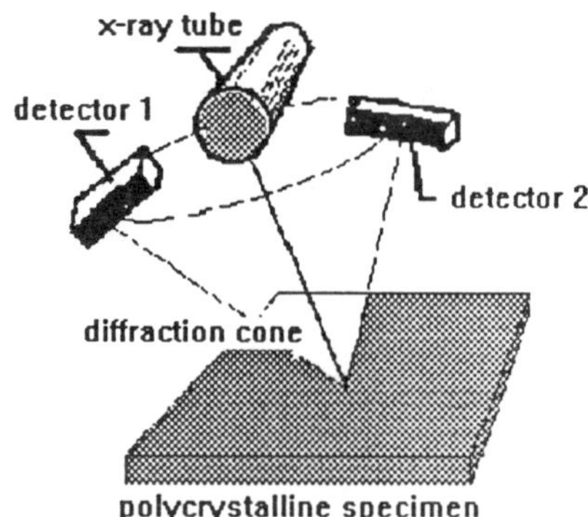

FIGURE 2 - HERE IS A TYPICAL DIFFRACTOMETER SETUP FOR STRESS MEASUREMENT WITH THE DIFFRACTION CONE OR DEBYE RING SHOWN. THE SOLID ANGLE OF THE RING IS DETERMINED BY THE BRAGG ANGLE OF THE DIFFRACTION PLANE.

Residual stresses are measured using either of two techniques. The first is the single exposure technique (SET), whereby a stress measurement is performed using only one beta β angle or tilt. This technique gives the user a very quick and efficient method to perform a stress measurement and is particularly suited for the need to take many measurements very quickly. The second is the multiple exposure technique (MET), whereby

multiple beta tilts are used in the analysis. This method is more revealing for materials for which the d vs. $\sin^2\psi$ relationship is not linear, as assumed in equation (3), but takes much longer than the SET (Klug and Alexander, 1974).

The stress measurements described in this paper were performed on steel using both the SET and MET with Cr $K\alpha_1$ radiation and the (211) hkl plane. The experimental error for each stress measurement ranged from 2 to 3 ksi.

A standard formula for calculating stress in thin walled pressure vessels (Faires, 1965) is used for theoretical stress calculations:

$$\sigma_h = \frac{P \times d}{2 \times t} \quad\dots\dots\dots\dots\dots(4)$$

Where: P is the pressure, d is the diameter of the cross section, t is the wall thickness and σ_h is the hoop stress. In this case, equation (4) represents a very simple mathematical model but is sufficiently accurate for the purposes of the stress measurements described henceforth..

FIELD SETUP AND EQUIPMENT

An x-ray diffractometer designed specifically for field stress measurements was used to perform measurements on a natural gas pipeline at two different excavation sites.

FIGURE 3 - THE MODIFIED DIFFRACTION HEAD IS SHOWN HERE ON THE PIPELINE COLLECTING DATA. THE POWER SUPPLY AND CONTROL HARDWARE ARE SHOWN IN THE FOREGROUND.

The first site was near a pumping station which had been excavated a number of times recently. The second site had not been excavated since the pipeline's installation some 40 years previously.

Since this particular section of the pipeline was relatively close to the San Andreas fault, it was suspected that the section of pipe at the unexcavated site may be under load due to ground settling and movement over the years. The section that had been recently excavated was suspected not to have a high degree of bending stresses. This is because some relaxation is assumed to have occurred when the pipeline was exposed and replaced with a new section. This of course is only speculation and it is with the x-ray diffraction measurement results that these questions and others can be answered.

The diffractometer is designed to measure stress at any point around the circumference of the pipe as long as that surface is exposed. It can also measure stresses in both the longitudinal and circumferential directions.

A portable electropolisher was used to prepare the surface at some of measurement locations. Electropolishing techniques are used because there is no other effective way to remove layers without introducing stresses.

Electropolishing techniques are essential in characterizing the subsurface stress gradients that can exist in pipeline materials.

FIGURE 4 - THE PORTABLE ELECTROPOLISHER IS SHOWN HERE. IT IS USED TO REMOVE MATERIAL LAYERS NONDESTRUCTIVELY.

DEPRESSURIZING GAS PIPELINE STRESS

Stress measurements were performed in the field on a 35 inch diameter steel gas pipeline with a wall thickness of 0.5 inches. The working pressure of this pipeline is usually 650 psig, but the maximum pressure used in these particular field measurements was 350 psig.

The site near the pumping station was excavated once again and the total stress (residual stress + applied stress) was subsequently characterized at several locations.

Some locations were electropolished to 0.003 and 0.005 inches below the surface to determine if there were any large stress gradients with depth. There was a small difference in stress with a maximum of about 3 ksi from surface to depth at the locations measured.

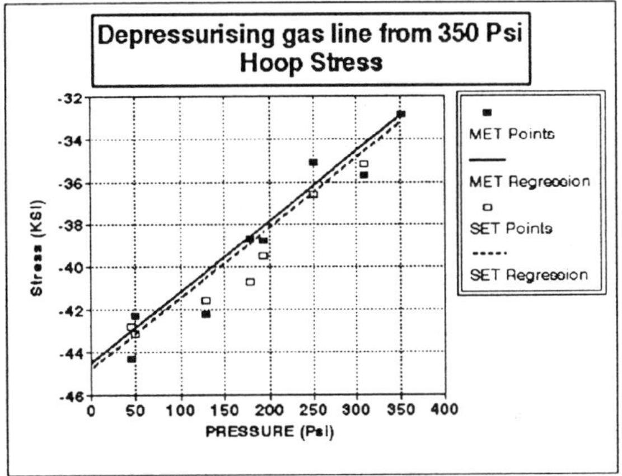

FIGURE 5 - THE REDUCTION IN TENSILE HOOP STRESSES APPLIED BY THE GAS PRESSURE INSIDE THE PIPELINE CAN BE OBSERVED IN THIS GRAPH.

Once the total stress was characterized at various locations, the pipeline was slowly depressurized. The hoop stress on the pipeline was monitored during the depressurization and a stress vs. pressure plot was created (see Figure 5.)

The net change in stress from 350 psig to 0 psig was determined to be 12.11 +/- 1.34 ksi using the MET and 11.59 +/- 1.16 ksi using the SET. These results were in close agreement with the expected change in stress of 12.25 ksi calculated using equation (4).

RESIDUAL STRESSES

To characterize residual stresses, longitudinal stress measurements were performed around the circumference of the section of pipe that had not been excavated since it's installation.

The results obtained are only a very small sampling used to paint a crude picture of the varying residual stresses around the pipeline. Ideally a more extensive detailed stress map would have been collected.

TABLE 1 - STRESS MEASUREMENT RESULTS.

Point	HOOP STRESS		LONGITUDINAL STRESS	
	(ksi)	(MPa)	(ksi)	(MPa)
1	-		-28	-191
2	-45	-311	-51	-349
3	-		-33	-231

The large differences in stresses measured from point to point (see results defined in Table 1) indicate that there is a large variation in the stresses around the circumference of the pipeline. The source of this variation in stress is unknown at this time and such a determination would require further investigation. These variations in the absolute stress may be due to ground movement since there was ample opportunity for this to occur over the pipeline's service life of more than 40 years.

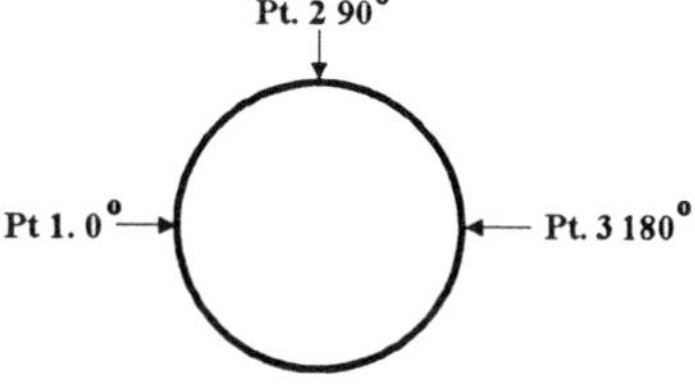

FIGURE 6 - LONGITUDINAL STRESS MEASUREMENTS WERE PERFORMED AT THREE DIFFERENT LOCATIONS AROUND THE CIRCUMFERENCE OF THE PIPE AT 90 DEGREE INTERVALS TO CHARACTERIZE VARIATIONS IN STRESS.

FOUR POINT BEND TEST

To increase the level of confidence in the x-ray method and the ability to measure stress on the original pipeline surface, without the need to electropolish, a four point bend test was performed on a small coupon cut from an original section of pipe removed from the pipeline.

This experimental technique has two specific purposes. The first being to determine if there is a linear response in the elastic region of the stress/strain curve plotted from the experimental data. It can clearly be seen that the plot in Figure 7 is indeed linear (correlation of .97784) when the coupon is loaded elastically.

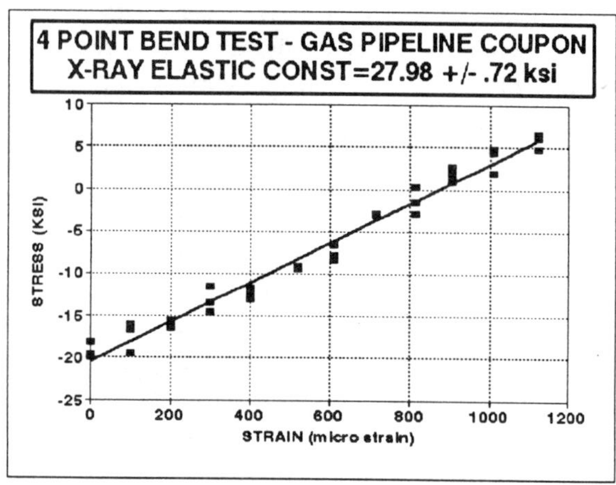

FIGURE 7 - PLOT OF STRESS VS. STAIN WITH STRESS MEASURED ON THE TOP OF THE COUPON USING X-RAY DIFFRACTION TECHNIQUES AND STRAIN MEASURED ON THE BOTTOM OF THE COUPON WITH A STRAIN GAGE.

This indicates that the light cleaning of the surface was an acceptable surface preparation technique and surfaces similar to this one should yield good results. The second purpose is to calibrate the x-ray elastic constant. In this case the x-ray elastic constant was determined and the experimental data was then recalculated on the basis of this slightly adjusted x-ray elastic constant.

CONCLUSIONS

Accurate absolute stress measurements can be performed on steel natural gas pipelines via x-ray diffraction stress measurement techniques.

Total stress measurements performed when the pipeline is at working pressure can be used to determine the state of stress under working loads. The stresses due to thermal expansion, fabrication, shot blasting or peening, and those stresses due to the pipe bending under its own weight and earth movement can also be easily characterized once the pipeline has been depressurized.

The stress measurement results from the pipeline depressurization experiment displays the x-ray diffraction method's ability to detect changes in total stress. Furthermore, the experimental results obtained exhibit excellent agreement with theoretical calculations.

In characterizing the residual stresses, it was found that the high variance in the measured stress values collected at 90 degree intervals around the pipe indicate that earth shifting may be a significant contributing factor to stresses recorded at these locations. Further investigation would be required to confirm this possible explanation for the large differences observed in some of the longitudinal stress measurement results.

Finally, the four point bend test performed indicates that the surface preparation techniques used were acceptable and that the x-ray elastic constant required a small adjustment.

The preceding evidence reinforces the suggestion that it is possible to characterize stresses in natural gas pipelines using the techniques of x-ray diffraction with a high degree of confidence. It also highlights the inherent advantages of the x-ray method.

EQUIPMENT

All measurements were performed using a **Proto XRD-1000©** diffraction head powered by the **Proto iXRD©** portable x-ray diffraction unit.

ACKNOWLEDGMENTS

The authors would like to thank Bill Amend of the Southern California Gas Co. for arranging and managing this experiment in a smooth and efficient manner.

REFERENCES

B.D. Cullity, 1978, "The Elements of X-Ray Diffraction", 2nd Edition, Addison-Wesley, Reading, Massachusetts.

V.M. Faires, 1965, "Design of Machine Elements", 4th Edition, The Macmillan Company, Collier-Macmillan Limited, Toronto, Ontario.

M.E. Hilley et.al., 1971, "Residual Stress by X-Ray Diffraction - SAE J784a", Section 2.2, Society of Automotive Engineers, Inc., Warrendale, PA.

H.P. Klug, L.E. Alexander, 1974, "X-Ray Diffraction Procedures", 2nd Edition, Wiley-Interscience, U.S.A.

I.C. Noyan, J.B. Cohen, 1987, "Residual Stress Measurement by Diffraction and Interpretation", Springer-Verlag, New York.

NDE-Vol. 13, NDE for the Energy Industry
ASME 1995

RESIDUAL STRESSES IN STEEL AND ZIRCONIUM WELDMENTS

John H. Root, Christopher E. Coleman and John W. Bowden
AECL Research
Chalk River, Ontario
Canada

Makoto Hayashi
Mechanical Engineering Research Laboratory
Hitachi, Ibaraki
Japan

ABSTRACT

Three dimensional scans of residual stress within intact weldments provide insight into the consequences of various welding techniques and stress-relieving procedures. The neutron diffraction method for non-destructive evaluation of residual stresses has been applied to a circumferential weld in a ferritic steel pipe of outer diameter 114 mm and thickness 8.6 mm. The maximum tensile stresses, 250 MPa in the hoop direction, are found at mid-thickness of the fusion zone. The residual stresses approach zero within 20 mm from the weld centre. The residual stresses caused by welding zirconium alloy components are partially to blame for failures due to delayed-hydride cracking. Neutron diffraction measurements in a GTA-welded Zr-2.5Nb plate have shown that heat treatment at 530°C for 1 h reduces the longitudinal residual strain by 60%. Neutron diffraction has also been used to scan the residual stresses near circumferential electron beam welds in irradiated and unirradiated Zr-2.5Nb pressure tubes. The residual stresses due to electron beam welding appear to be lower than 130 MPa, even in the as-welded state. No significant changes occur in the residual stress pattern of the electron-beam welded tube, during a prolonged exposure to thermal neutrons and the temperatures typical of an operating nuclear reactor.

INTRODUCTION

Neutrons penetrate easily through many engineering materials and the three-dimensional mapping of the residual stress state in the interior of weldments is straightforward. The measured patterns of residual stress provide knowledge of the probable positions for fracture, the effectiveness of stress-relieving treatments and the accuracy of finite-element calculations of the welding process. Weldments of ferritic steel, stainless steel, and of nickel, zirconium and titanium alloys have all been studied by the neutron diffraction stress-scanning method (Mahin et al., 1991, Jo et al, 1993, Root et al. 1993, Coleman et al., 1993). In this paper, the principles of the method are explained, and examples are presented of residual stress distributions in weldments of ferritic steel and zirconium alloys. The zirconium alloy weldments are particularly interesting because, like titanium, zirconium has a hexagonal close-packed crystal (hcp) structure. During the welding cycle, a sequence of phase transformations re-orients the crystallites in the heat-affected zone, which can increase the susceptibility of the material to delayed hydride cracking.

NEUTRON DIFFRACTION MEASUREMENTS

Neutron diffraction is a method for measuring the spacing, d, between the atomic planes of a crystal lattice. A neutron beam of known wavelength, λ, is diffracted from its incident direction by a scattering angle, 2θ, according to Bragg's law,

$$\lambda = 2d \sin\theta \qquad (1).$$

By scanning a detector through a range of scattering angle, a profile of neutron counts versus 2θ is obtained, as shown in Fig. 1. A Gaussian function is fitted to the

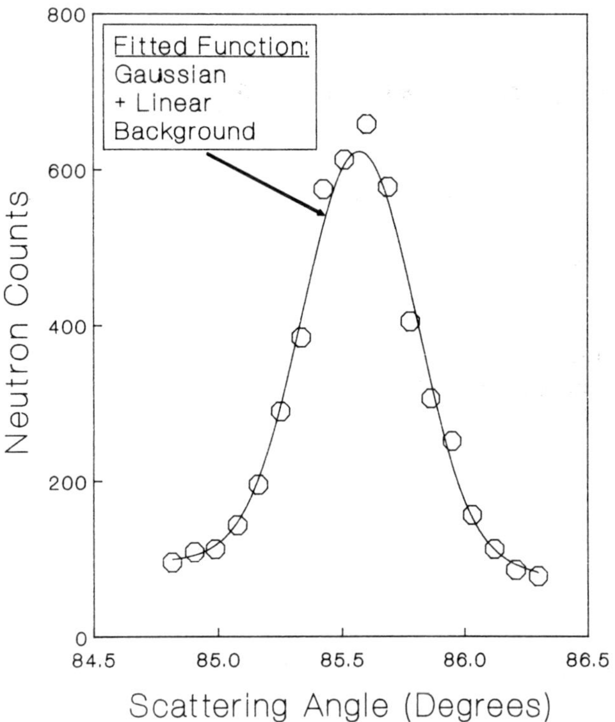

FIG. 1 - A NEUTRON DIFFRACTION PEAK

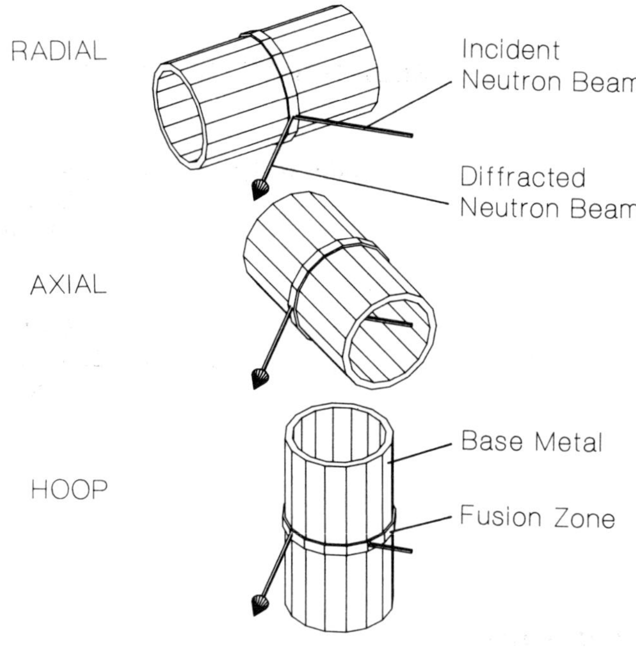

FIG. 2 - ORIENTATIONS TO MEASURE STRAIN

raw data to obtain the mean scattering angle to a typical precision of $\pm 0.003°$. Values of d in most engineering materials range from 0.1 nm to 0.3 nm and are determined through Eq. (1) to a precision of about $\pm 1 \times 10^{-5}$ nm. Elastic strain, ε, is determined by comparing the measured value of d to the value measured in a suitable stress-free reference, d_o through the relation,

$$\varepsilon = d/d_o - 1 \qquad (2).$$

Strain components are measured in a direction that bisects the incident and diffracted neutron beams. To obtain values of strain in the principal directions of a tubular weldment, axial (A), radial (R) and hoop (H), the tube must be reoriented to place this bisector in the required direction, as shown in Fig. 2. The incident and diffracted beams are shaped by slits in neutron-absorbing cadmium masks. Measurements are made only in the region of intersection of these two beams, which is called the sampling volume. The weldment is moved with a computer-controlled XYZ translator, to place the sampling volume at a selection of locations within the weldment. In this way, a spatial map of strain is constructed.

With three measurements of strain at each location, the residual stresses can be calculated through a generalized Hook's law,

$$\sigma_A = [E/(1+\nu)] \, [\varepsilon_A + (\nu/(1-2\nu)) \, (\varepsilon_A + \varepsilon_H + \varepsilon_R)] \qquad (3),$$

where E is Young's modulus, ν is Poisson's ratio, and the subscripts indicate the components of strain and stress.

BUTT WELD IN A FERRITIC STEEL TUBE

The residual stresses near a circumferential butt-weld in a ferritic steel tube were examined by the neutron diffraction method. The weldment was not stress-relieved. The tube had an outer diameter of 114 mm and a wall thickness of 8.6 mm. For measurements of the axial and radial strain components, the sampling volume had dimensions of 1.5 mm x 1.5 mm x 10 mm, with the long dimension tangent to the curvature of the tube. For measurements of the hoop component of strain, the height of the sampling volume was reduced to 2 mm to retain sufficient spatial resolution in the axial direction of the tube to scan the gradient of strain with distance from the centre of the weld.

The residual stresses are created primarily by the shrinkage of the fusion and heat-affected zones on cooling. It is easy to visualize shrinkage in the hoop direction. However, the fusion and heat-affected zones cannot shrink completely because they are constrained by the material in the remainder of the tube. A tensile hoop component of residual stress develops near the weld,

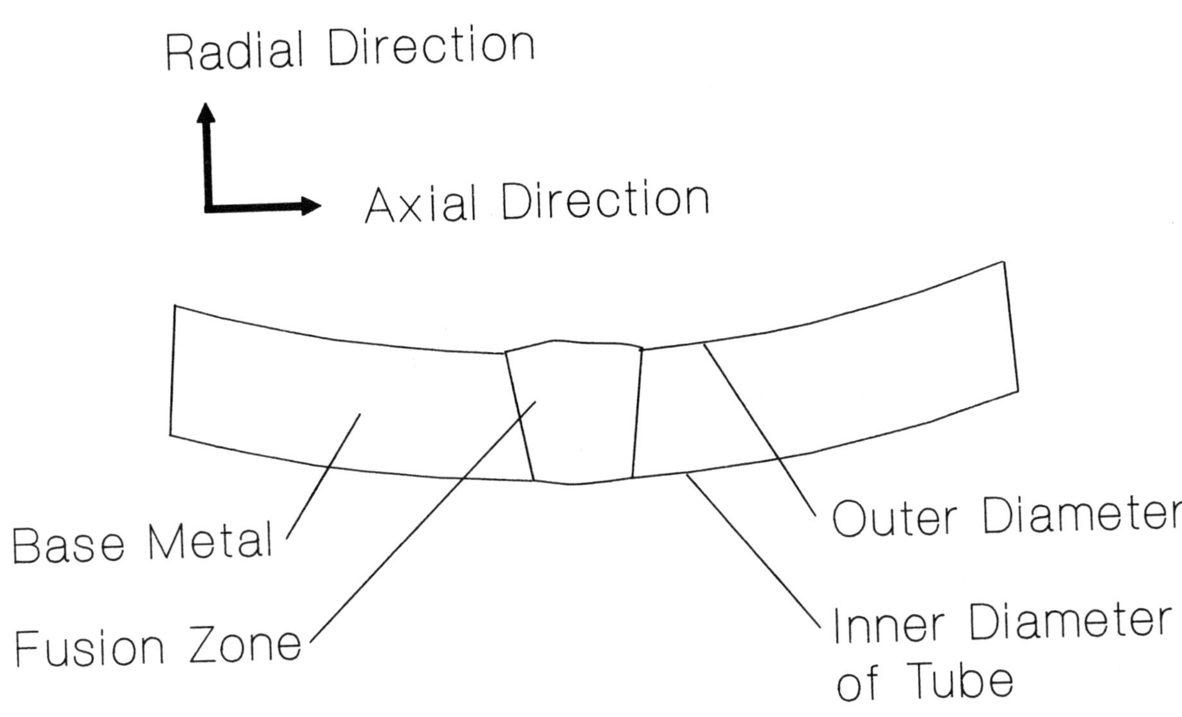

FIG. 3 - DISTORTION OF THE TUBE WALL CREATES AXIAL RESIDUAL STRESSES

balanced by compression in the base metal of the tube, far from the weld. As the weld shrinks in the hoop direction, a secondary effect develops in which the tube wall is distorted by an inward-directed radial force. The wall of the tube bends in the axial-radial plane, as depicted in Fig. 3. The geometry is reminiscent of a three-point bend, and one expects to find tension in the axial direction near the inner diameter of the tube, and a corresponding axial compression near the outer diameter of the tube.

In Fig. 4, the measured residual stresses are shown throughout a section of the weldment that extends from the inner to the outer diameter of the tube, and from the weld centre to a distance of 30 mm in the axial direction. The dominant residual stress is in the hoop direction. The maximum tensile hoop stress, (250 ± 30) MPa is located at mid-thickness of the fusion zone. This value of stress is close to the yield point of the material. The hoop stresses approach zero within 20 mm from the weld centre. Close to the weld, the axial stress is compressive near the outer diameter and slightly tensile near the inner diameter of the tube. If the axial stresses were created only by the mechanism shown in Fig. 3, the compressive stresses are expected to be similar in magnitude to the tensile stresses. However, the compression near the inside diameter is greater than the tension near the outside, so additional mechanisms contribute to the net residual axial stress pattern. For example, by the time the inner region of the tube cools, the outer region may already have yielded plastically and hardened. Then, as the inner region contracts, it adds compressive axial stresses to the outer region of the tube. This spatial-dependence of cooling rate should likewise enhance the compressive hoop

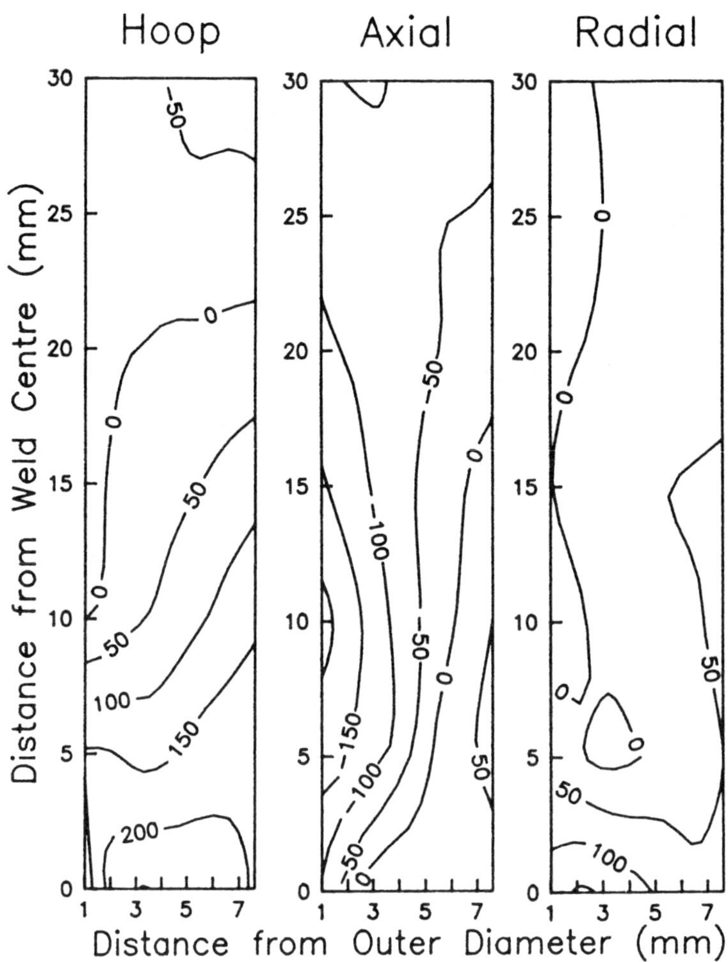

FIG. 4 - RESIDUAL STRESSES (MPa) NEAR A BUTT-WELDED FERRITIC STEEL TUBE

stresses near the outer diameter, as is observed.

While the various mechanisms that create residual stresses in the butt weld may be related in a complicated way, there is no doubt that tensile axial and hoop stresses make the inner surface of a tube more susceptible to chemical corrosion and stress-corrosion cracking may shorten the lifetime of the weldment.

HEAT TREATMENT OF A GTA WELD IN A Zr-2.5Nb PLATE

A gas-tungsten-arc (GTA) weld joined two Zr-2.5Nb plates that were made by hot-rolling at 750°C and annealed at 730°C for 1 h prior to welding. The plates were 6 mm thick, 190 mm wide and 300 mm long. The weld line was parallel to the transverse direction of the original rolled plates. The rolling texture of the base metal exhibited a high concentration of basal poles close to the transverse direction. In the heat-affected zone (HAZ) of the weld, the thermal cycle of welding brings the material from the low-temperature (hcp) α phase to the high-temperature, body-centred cubic β phase. On cooling, the material re-transforms to the α phase, but there are 12 distinct variants for this transformation, and the resulting texture has a large volume fraction of basal

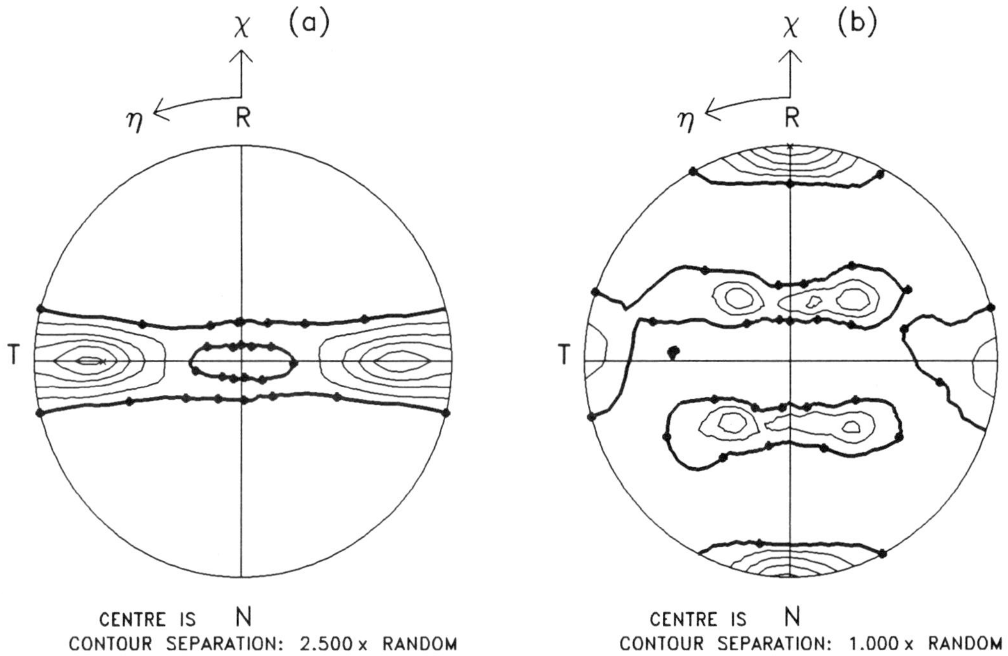

FIG. 5 - BASAL POLE FIGURES IN (a) THE BASE METAL (b) THE HEAT-AFFECTED ZONE OF A WELDED Zr-2.5Nb PLATE

poles perpendicular to the line of the weld, in the plane of the plate. The basal pole figures in the base metal and HAZ are shown in Fig. 5. The texture of the HAZ is very unfavourable for DHC susceptibility because hydrides tend to form as platelets that are parallel to the basal planes of the α phase. If the concentration of hydrogen in the metal is high enough to cause precipitation of hydrides on cooling, many platelets will form with their planes extending parallel to the weld and through the thickness of the plate. If, in addition, there are significant tensile stresses, the susceptibility of the HAZ to delayed hydride cracking will be increased.

Scans of the residual strain parallel to the line of the weld are presented in Fig. 6. The primary stress in a linear weld is created by the shrinkage of the fusion and heat-affected zones on cooling. The shrinkage is constrained by the material in the base metal of the weldment. Residual tension develops parallel to the weld, near the fusion zone. This stress is balanced by compressive longitudinal stresses far from the fusion zone. The longitudinal strains reflect this underlying residual stress state, with positive, tensile values close to the weld and negative, compressive values that reach a minimum

30 mm from the weld centre. In Fig. 6, a comparison is made between the longitudinal strain patterns in the as-welded state and in the state following a stress-relieving heat treatment of 530°C for 1h. The maximum value of strain is reduced by about 60% after the heat treatment.

Experience has shown that GTA welds like this one are very susceptible to DHC failures if they are not stress-relieved shortly after the weld is made. However, stress relief by heat treatment has been effective in eliminating such failures (Coleman et al., 1993). Clearly, the heat treatment serves to reduce the residual stress sufficiently to offset the enhancement of DHC susceptibility that is caused by the texture in the HAZ. Apparently, the observed, partial reduction of the residual stress is sufficient to achieve the desired reliability of the weld.

AGING OF AN ELECTRON BEAM WELD IN A NUCLEAR REACTOR

In contrast to the behaviour of large GTA welds in zirconium alloy components, electron-beam (EB) welds in Zr-2.5Nb pressure tubes have rarely failed (Coleman et al., 1993). Two EB weldments in Zr-2.5Nb pressure tubes have been compared. The first weldment was

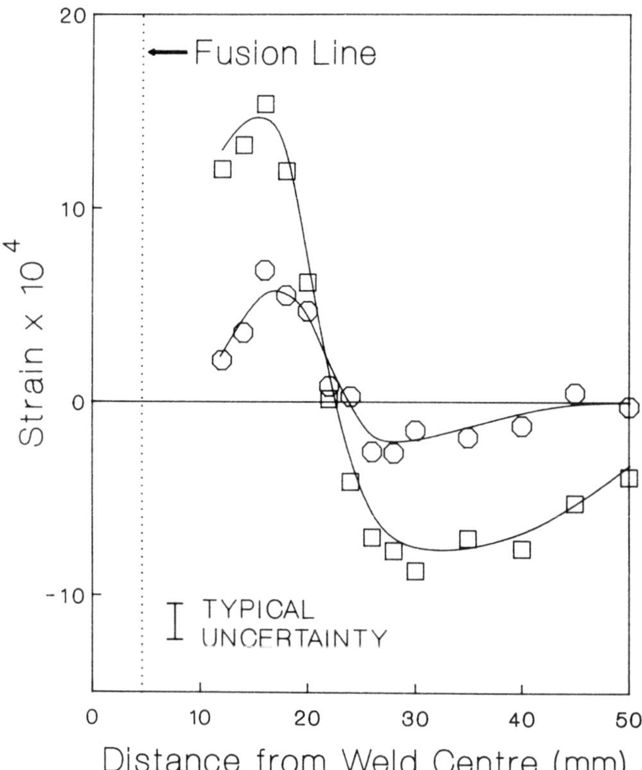

FIG. 6 - STRAIN PROFILES AS-WELDED (□) AND HEAT-TREATED (○) Zr-2.5Nb

examined by neutron diffraction shortly after its fabrication, to determine the residual strain pattern in the as-welded condition. The second, was in service for eight years in an experimental loop in the NRU nuclear research reactor at Chalk River Laboratories. While in service, this weldment was exposed to a field of thermal neutrons and temperatures ranging between 250°C and 300°C. Because of its exposure to neutrons, the weldment was radioactive and emitted gamma radiation at levels as high as 25 R/h on contact. To obtain a scan of the residual strains in this weldment, a special shield was needed to permit safe handling on the neutron diffractometer. A lead annulus of thickness 60 mm, supported by a steel case, reduced the fields to an acceptable level. Small windows through the lead annulus allowed the incident and diffracted neutron beams to access the weldment inside the shield (Fig. 7).

The primary residual stress in these EB welds is in the hoop direction, created by the constrained shrinkage of the fusion and heat-affected zones on cooling from the welding temperature. As in the GTA weld above, there was a texture re-orientation in the HAZ. The textures and microstructures of the HAZ and fusion zone were such that reliable diffraction-peak data could not be obtained in the hoop direction, so residual stresses could only be determined at distances greater than 5 mm from the centre of the weld. However, complete patterns of the axial and radial strains showed no increase in strain magnitude on closer approach to the weld centre. Therefore it can be concluded that the residual stresses do not exceed the values observed, even in the fusion zone of the weldment.

The variation of hoop stress with distance from the weld centre is compared for both weldments in Fig. 8. The now-familiar pattern of a tensile stress that increases on approaching the weld centre is observed in both weldments. Small differences between the patterns of the new and aged weldments may be due to variability from one weld to the next, rather than the effects of the prolonged exposure of the older weld to radiation and elevated temperature. Even in the new weldment, which was not stress-relieved, the residual hoop stress is less than 130 MPa. This welding residual stress is much lower than the 0.2%-offset yield stress of the material, which is typically greater than 300 MPa. Perhaps the historical reliability of electron-beam welds in zirconium alloys may be traced to the low levels of residual tensile stress created by the joining process.

CONCLUSIONS

Neutron diffraction provides detailed spatial maps of the residual stress and strain fields within weldments. The maps pinpoint regions where stresses are near the yield point of the material and provide evidence of the effectiveness of stress-relieving heat treatments. The good performance of GTA welds in Zr-2.5Nb, which have been heat treated at 530°C for 1 h, can be traced to a partial reduction of residual stresses, compared to those in the as-welded condition. The historical good performance of electron beam welds in pressure tubes in the NRU reactor may be traced to the low residual stresses present in as-welded tubes. Aging of these weldments for up to eight years at reactor operating temperatures and in a thermal neutron flux produces no significant effect on the pattern of residual stresses.

ACKNOWLEDGEMENTS

This work was funded by the Mechanical Engineering Research Laboratory of Hitachi, Ltd., by AECL Research, Chalk River and by the CANDU Owners Group, WPIR 3326, Working Party 33. The specimen container for the radioactive weldment was developed by S. Donohue of the Reactor Materials Research Branch at Chalk River.

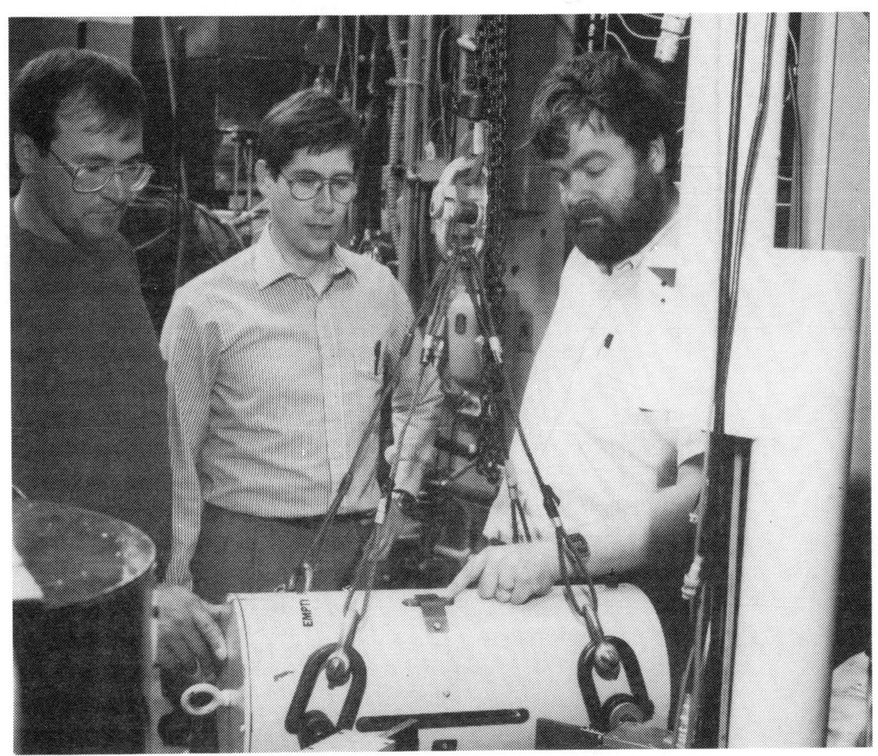

FIG. 7 - SPECIMEN SHIELD FOR THE IRRADIATED WELDMENT

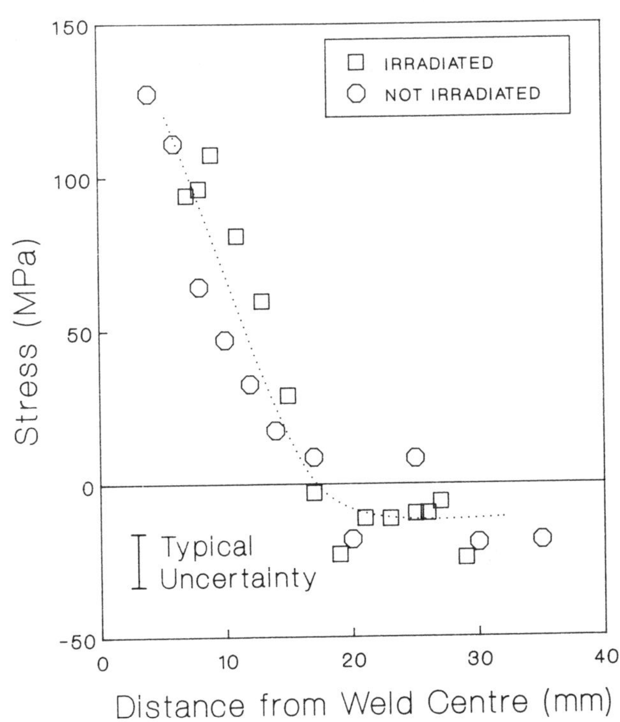

FIG. 8 - HOOP STRESS IN IRRADIATED (□) and
UNIRRADIATED (○) WELDMENTS

REFERENCES

Mahin, K.W., Winters, W.S., Holden, T. and Root, J., 1991, "Measurement and Prediction of Residual Elastic Strain Distributions in Stationary and Travelling Gas Tungsten Arc Welds", in *Practical Applications of Residual Stress Technology*, C.O. Ruud, ed., ASM International, Materials Park, OH.

Jo, J., Wang, X.L., Kleinosky, M.J., Green, R.S., Hubbard, C.R. and Spooner, S., 1993, "Evaluation of Stress Relief Treatments Using Diffraction", *Proc. 4th Int. Conf. on Residual Stresses*, S.A. David and J.M. Vitek, ed., ASM International, Materials Park, OH.

Root, J.H., Holden, T.M., Schroeder, J., Hubbard, C.R., Spooner, S., Dodson, T.A. and David, S.A., 1993, "Residual Stress Mapping in a Multipass Ferritic Steel Weld", *Materials Science and Technology*, Vol. 9, pp. 754-759.

Coleman, C.E., Doubt, G.L., Fong, R.W.L., Root, J.H., Bowden, J.W., Sagat, S. and Webster, R.T., 1993, "Mitigation of Harmful Effects of Welds in Zirconium Alloy Components" *Proceedings of 10th Int. Symp. on Zirconium in the Nuclear Industry*, ASTM, Philadelphia PA 19103 (in press). Report AECL-10950, available from the Document Office, AECL Research, Chalk River, Ont. Canada, K0J 1J0.

DIFFRACTION MEASUREMENTS OF LINEWIDTH
IN PLASTICALLY DEFORMED AND FATIGUED HY–80 MATERIALS

T. M. Holden, John H. Root, and J. Fox
AECL Research
Chalk River, Ontario, Canada

J. Porter
Defense Research Establishment Atlantic
Halifax, Nova Scotia, Canada

James A. Pineault and Michael E. Brauss
Proto Manufacturing Limited
Oldcastle, Ontario, Canada

ABSTRACT

Neutron and X-ray diffraction measurements have been carried out on plastically deformed and fatigued samples of HY-80. The line broadening and intensity correlate well with the plastic deformation offering the possibility of a non-destructive probe of plasticity. For loads in the range considered so far, there is no increase in width with the number of fatigue cycles. The method has been applied to measuring the plastic zone near the tip of cracks in HY-80 bars.

1. INTRODUCTION

Non-destructive methods of measuring plastic deformation or fatigue damage in marine components which have been subjected to stresses beyond the yield point or to cyclic loading are of importance to estimates of remaining life in naval vessels. X-ray diffraction offers the possibility of assessing plasticity and fatigue damage. With the available portable X-ray equipment, tests could be made in operating vessels on a routine basis. X-ray diffraction provides information about the first few microns below a surface whereas neutron diffraction gives this information as a function of depth. The two methods complement each other and provide a check on each other. The neutron and x-ray diffraction studies presented here give an excellent correlation between induced linewidth and plastic deformation. In addition it is also evident that the intensity of the diffraction peak depends

on the degree of plastic deformation, thus providing a more sensitive means of non-destructive evaluation of plastic deformation. On the other hand, there is no evidence that the intensity, width or position of diffraction peaks, in either X-ray or neutron experiments are sensitive to fatigue damage in the range of loads investigated here.

The physical origins of line-broadening have been known for decades and were, for example, discussed by Cullity [1] in 1956. Deformation generates dislocations within individual polycrystalline grains, which permit a non-uniform strain distribution to exist within a grain and which can break up the coherent scattering from the whole grain into smaller sub-grains. These two effects contribute differently to the experimental lineshape [2] and have different variations with scattering angle. The broadening due to size, B, is given by "the Scherrer formula" [1]

$$B = \frac{0.9\lambda}{t\cos\theta} \qquad (1)$$

where B is the full width at half maximum in radians, λ is the wavelength, t is the diameter of the diffracting volume and 2θ is the scattering angle. The broadening due to strain, b, is related [1] to the full width at half $\Delta d/d$ height of the distribution of plane spacing about the average value by

$$b = - 2\frac{\Delta d}{d}\tan\theta \qquad (2)$$

A thorough discussion of extracting these two effects from experimental data has been given by Langford [2]. We have taken a very elementary approach in this paper for extracting the effects of plastic deformation which assumes that both the intrinsic linewidth and the instrumental linewidth are Gaussian in form.

This paper is arranged as follows. Section 2 describes the neutron and X-ray diffraction experiments and the basic theory required to derive the results. The experimental results for plastically deformed samples are presented and discussed in section 3. Section 4 is devoted to measurements near a crack tip in a bar of HY-80.

2. EXPERIMENTS

The intensity measured in a diffraction experiment has the general form of a sharp peak due to the coherently scattered radiation on a slowly varying background. The coherently scattered radiation carries the information required about lattice spacing, grain orientation and deformation from the peak position, $2\theta_{hkl}$, the integrated intensity I_{hkl} and the full-width at half-maximum, Δ_{hkl}. The parameters 2θ, I, Δ were obtained by fitting a Gaussian plus a sloping background to the data in the form of intensity versus 2θ. For X-ray diffraction, the line-profile is a double-peak shape [1] because of the closely spaced $K\alpha_1$ and $K\alpha_2$ lines of the characteristic X-ray spectrum. The width of the peak can be characterized by the full-width at half-maximum or alternatively by the integral breadth, that is the ratio of the integrated intensity to the peak height.

The spread of the wavelength of radiation and the angular collimation of the incident and scattered beams contribute to an instrumental linewidth Δ_I for both X-rays and neutrons. In order to maximize the accuracy of the part of the linewidth originating in the deformation, the instrumental linewidth must be minimized by careful equipment design. If we assume that the intrinsic deformation lineshape and the instrumental lineshape are both Gaussian, then they are combined as follows

$$\Delta^2 = \Delta^2_I + \Delta^2_D \qquad (3)$$

2.1 Neutron Diffraction

The experiments were carried out with a high resolution arrangement with the (331) planes of a Silicon monochromator selecting neutrons of wavelength 1.9740Å at a take-off angle of $2\theta_m = 105.0°$. The angular collimations between the monochromator and sample, and the sample and the detector were 0.16° and 0.125° respectively. Accurate linewidth measurements require high statistical definition of the peak and typically the peak intensity was about 5000 counts in 1 hr. The wavelength of the neutron beam was calibrated with a standard Ni powder. By chance the (111) powder peak of Ni is very close in angle to the (110) peak in HY-80 steel and the former gives a measure of the instrumental contribution to the linewidth. Two separate calibration experiments gave the widths to be $0.291 \pm 0.005°$ and $0.283 \pm 0.003°$ for Δ_I. The experiments were carried out in transmission geometry on the (110) peak of HY-80 steel dogbone samples mounted on a carousel so that six samples could be measured consecutively. Some checks were also made of the samples in reflection geometry and there was no change in the widths. Many measurements were made on fatigue samples which all showed widths of 0.282 $\pm 0.003°$. This was taken as the definitive measure of Δ_I.

2.2 X-ray Diffraction
The experiments were carried out with the XRD-1000 Proto X-ray spectrometer with the $K\alpha_1$ (1.9399Å), $K\alpha_2$ (1.9360Å) characteristic doublet of Iron. The collimation before the sample, governed by the slit size near the sample (2 mm) and the aperture near the tube (5 mm), separated by 40 mm was 5.0°. The collimation after the sample, governed by the spot size at the sample and the 0.025 mm separation of detecting elements was 1.5°. The plane used was the (110) plane of the body-centred cubic structure and the scattering angle was about 58°. Thus both the wavelength and scattering angle were nearly the same for the X-ray and neutron measurements. The X-ray measurements were made in reflection geometry.

2.3 Samples.
Standard "dogbone" samples 0.18 in. (4.9 mm) thick, 1.0 in. (25.4 mm) wide and 7.5 in. (190.5 mm) long were machined from HY-80 stock. The sample surfaces were electro-etched with a grid of circles. The samples were then plastically deformed to different levels. The surface plastic strains which vary along the length of the test samples were obtained by measuring the now elliptical etch patterns with the FAMOSS system. The engineering strain was converted to true or logarithmic strain. The X-ray spot size was of order of the etch pattern size: the volume irradiated with neutrons was pillar-shaped of width 2 mm and height 5 mm, thus averaging over the middle 2 etch patterns of

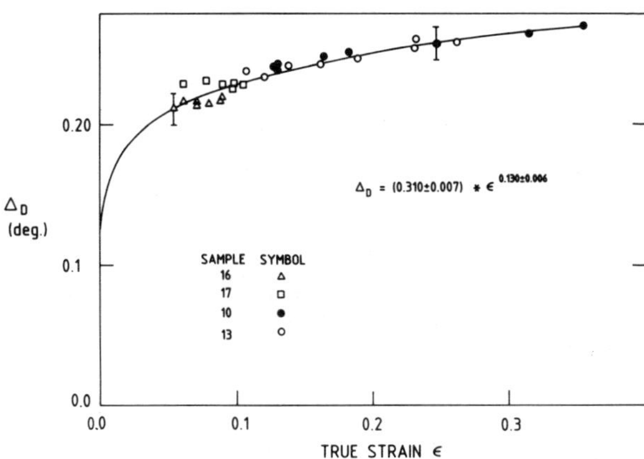

Fig. 1.   The variation of the intrinsic linewidth associated with deformation as a function of the true strain.

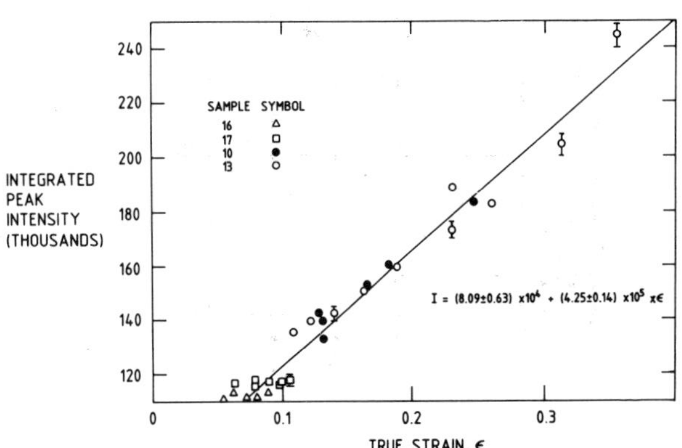

Fig. 2.   The variation of the integrated intensity of the (110) diffraction peak as a function of the true strain.

which there were 6 across the width.   The neutron measurements were made at mid-thickness of the sample whereas the X-ray measurements were made at the surface.

3. EXPERIMENTAL RESULTS

The intrinsic linewidth, due to deformation, for all the samples, calculated from the measured linewidth and an instrumental linewidth of $0.282 \pm 0.003°$ with the aid of eq. 3, is shown in Figure 1 plotted versus true strain.   The solid curve fitted to the data is

$$\Delta_D = K \, \varepsilon^n \qquad (4)$$

with $K = 0.310 \pm 0.007°$ and $n = 0.130 \pm 0.006$.

The intensity, corrected for transmission through the samples, is plotted versus true strain in Figure 2.   The dependence of intensity with plastic strain is linear in the region studied showing that the intensity change of the diffraction peaks can be used to measure plasticity and is much more sensitive than the linewidth.   The intensity followed the relation

$$I = (8.09 \pm 0.63) \times 10^4 + (4.25 \pm 0.14) \times 10^5 \, \varepsilon. \qquad (5)$$

It has long been recognized that, in tension, the <110> axes of b.c.c. materials rotate towards the tensile axis [3], since this is the effect in single crystal studies, and the end texture for the case of constrained tension, such as, wire drawing or extrusion.

Figure 3 shows the variation of the integral breadth of the (110) diffraction peak as a function of position along the length of the samples as determined by X-ray diffraction.   For a Gaussian peak, the integral breadth should be 6% wider than the full width at half maximum. Thus the X-ray instrumental linewidth is about x2 that of the neutron instrumental linewidth and yet the effects of plasticity are readily seen.   Further reductions in X-ray linewidth by modification of the X-ray optics are planned.

Contrary to sizeable effects noted for plastically deformed samples, there was no change in the linewidth or intensity as a function of number of fatigue cycles at a maximum stress of 550 MPa (which is close to the yield stress of HY-80).

4. MEASUREMENTS OF LINEWIDTH NEAR A CRACK-TIP

The measurements were made at a neutron wavelength of 2.4824 Å, reflected from the (113) planes of a Ge monochromator, giving a diffraction peak near a scattering angle of 75°.   The HY-80 samples were bars 254 mm long, 25.4 mm wide and 40 mm thick, with cracks which penetrated 6 mm and 20 mm into the thickness.   The samples were loaded up to 80% of the plastic load limit in 3-point bending.   The region over which the width was averaged was of dimensions 2x2x2 mm³ as defined by slits in absorbing cadmium placed in the incident and scattered beams.   The minimum width observed far from the crack was 0.44 ± 0.01 which is 50% greater than in the experiments previously described and is primarily due to worse instrumental resolution.

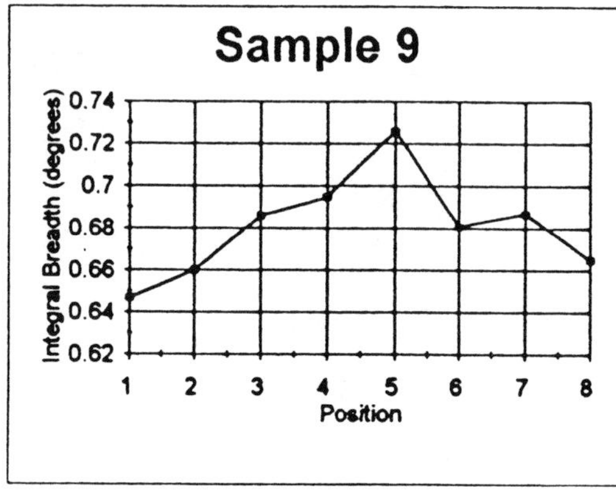

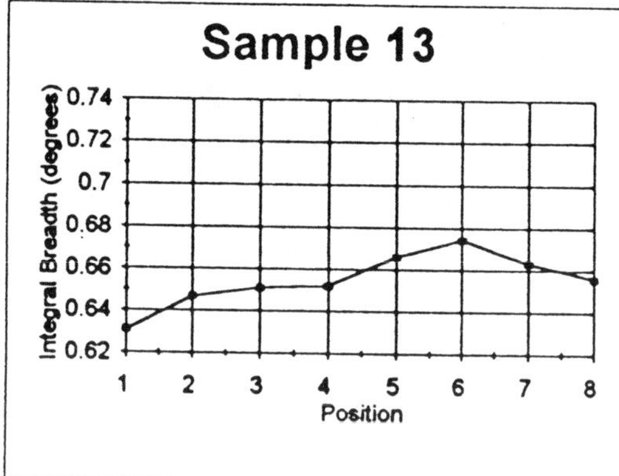

Fig. 3 X-ray measurements of linewidth for samples 9 and 13 as a function of position along the length of the sample.

The measured linewidths have a magnitude between 0.44° and 0.5° with a typical uncertainty of 3%. The results for the 6 mm crack are plotted as a function of position and load in Figure 4. The results for 20-50 mm from the crack show a significant thorough-thickness variation which is load independent and which probably originates in the manufacture of the bar. The curve drawn through the data for 20-50 mm is reproduced for all the other distances as a base-line against which to measure load-or crack-induced effects. There are no additional contributions to the linewidth until the offset from the crack is 8 mm or less. An intrinsic width is observed extending 12 mm in from the compressive face at a load of 83.4 kN (80% of the plastic load limit). At a lateral displacement from the crack centreline of 4 mm an intrinsic width which is independent of load is observed. On the centreline a 20% increase of linewidth is observed very close to the nominal position of the crack tip, and is independent of load. The intrinsic linewidth calculated with eq. 3 corresponding to the maximum increase is 0.29 ± 0.05° From Figure 1 this would correspond to a plastic strain of greater than 0.3. The width increase along the

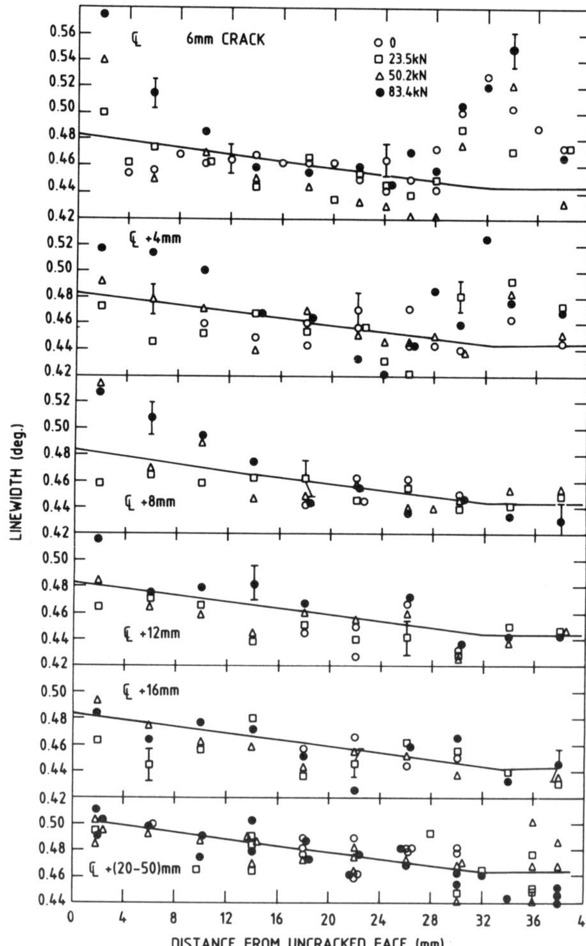

Fig. 4 Measured linewidths as a function of position for the shallow crack sample. The solid curve through the data is the best average width far from the crack. Enhanced linewidths are observed near the crack tip and opposite the tip on the compressive face.

centreline has disappeared 5 mm from the position of the crack tip. The region of enhanced width, a region with a radius of about 5 mm, is identified with the plastic zone. The size of the plastic zone may be calculated from linear elastic fracture mechanics [4] knowing the stress intensity factor for the geometry of the crack. In the plane stress approximation, i.e., near the surface, the plastic zone is estimated to be 3.8 mm, and in the plane strain approximation, i.e., at depth, the plastic zone should be three times smaller. Bearing in mind the size of the gauge volume the measurements are not out of line with the estimated value.

ACKNOWLEDGEMENTS

We wish to acknowledge the expert advice of R.R. Hosbons in the applications of fracture mechanics and the preparation of fatigue samples. L. McEwan, B.L. Wotton and J. Mecke provided invaluable technical assistance.

REFERENCES

1. B.D. Cullity, "Elements of X-ray Diffraction", (Addison-Wesley: Reading, Mass), 1956.

2. J.I. Langford in "Accuracy in Powder Diffraction II", Eds. E. Prince and J.K. Stalick, NIST special publication <u>846</u>, 173, 1992.

3. I.L. Dillamore and W.J. Roberts, Metallurgical Reviews <u>10</u>, 272, 1965.

4. S.T. Rolfe and J.M. Barsom, "Fracture and Fatigue Control in Structures" (Prentice-Hall, Inc.: New Jersey), 1977.

RESIDENTIAL STRAIN MEASUREMENTS OF STEEL STRUCTURES

Michael E. Brauss and James A. Pineault
Proto Manufacturing Limited
Oldcastle, Ontario, Canada

ABSTRACT

The following paper describes the results of strain measurements performed insitu on several steel structures using x-ray diffraction techniques. It will highlight examples where a truly portable x-ray diffraction instrument has been used to characterize residential (or absolute) stresses nondestructively in new and existing steel structures. The paper suggests new approaches to the characterization of the stress state in steel structures made possible by the technique itself.

INTRODUCTION

Understanding the actual stress states that reside in both new and existing steel structures has been an ongoing concern for engineers over the years. The requirement to extend the useful life of aging structures, as well as safety issues, has created a need for structural stress characterization to help optimize the allocation of limited maintenance dollars. The financial pressures to obtain the required performance and reliability in new structures with reduced material and fabrication costs have prevailed upon engineers to design closer and closer to what might be termed "safe" limits. "...Residual stresses are receiving increased attention by the engineering community...concerned with pressures to reduce the cost of materials used in structures, extending the useful lifetime of existing structures and the demand for greater reliability ..." (Ruud 1982). Although finite element modeling techniques are becoming more and more sophisticated, in many cases the stress state in an existing structure is too complex to be easily or economically modeled (Dunker and Rabbat, 1993) and the residual stresses present cannot be inferred. Thus, for the practical implementation of computer based stress modeling techniques, the residual stresses as well as selected applied stresses must be measured independently. "It is clear that engineers must know what residual stress is present, if any, before they can know the actual stress corresponding to a particular service load."
(Hilley et. al., 1971)

A means of measuring residual stress accurately and nondestructively in the laboratory has existed for some time. "A notable advance was made in 1953 when it was shown that residual stress could be measured quickly...by means of a new x-ray diffractometer" (Hilley et. al., 1971). However, advances in detector technology and system miniaturization have only recently made x-ray diffraction available as a versatile and practical tool for the nondestructive characterization of residual stress in the field. (Pineault and Brauss, 1994)

APPARATUS

An x-ray diffractometer designed specifically for field stress measurements was used to perform measurements on several different steel structures. In each case the equipment was hand carried by the operator and was set up at the various sites in 20 to 30 minutes. The diffraction head weighs about 12 kilograms and the power supply and control unit weighs in at about 18 kilograms. For measurements requiring material removal, a portable electropolisher weighing about 8 kilograms was used. Power was not readily available at all of the measurement sites thus a portable gasoline generator was employed as a source of A.C. power. A portable computer was used for data acquisition and analysis.

BASICS OF THE X-RAY DIFFRACTION STRESS MEASUREMENT TECHNIQUE

The x-ray method does not measure stress directly but measures strain from which stress values are calculated. The x-ray method rather elegantly takes advantage of the crystalline structure of the material itself, by using the atomic lattice spacing as a strain gage. As a result, thousands of "built in strain gages" within the crystals which compose the material are available for strain measurement by the x-ray diffraction method. To phrase it more exactly, the surface strain present can be determined by the measurement of the elastic atomic lattice spacing or "d-spacing" as it is commonly called. This lattice spacing, the distance between the planes of atoms, is a function of the material and the stresses present in the material. The x-ray diffraction angle θ for a given x-ray wavelength λ can be used to determine the material "d" spacing by solving Bragg's law:

$$n\lambda = 2d\sin\theta \dots\dots\dots(1)$$

For x-ray diffraction to occur, i.e. constructive wave interference, the path difference traveled by the diffracted beam through the material, as compared to a non-diffracted beam, must be equal to $n\lambda$ (Noyan and Cohen, 1987). The presence of residual stresses in the material produces a shift in the x-ray diffraction peak angular position (Cullity, 1978) which is directly measured by the detector.

Once the lattice d-spacings are measured for the unstressed (d_o) and stressed (d_i) material conditions, the atomic lattice strain can then be calculated by the following relationship (Hilley et. al., 1971):

$$\text{strain} = (d_i - d_o)/d_o \dots\dots\dots(2)$$

For isotropic materials, strains can be converted to stress values using the equation shown below.

$$\text{stress } (\sigma) = \frac{d_\psi - d_O}{d_O}\left(\frac{E}{1+\upsilon}\right)\frac{1}{\sin^2\psi} \dots\dots\dots(3)$$

where: $\frac{E}{(1+v)}$ is the x-ray elastic constant, ψ is the angle subtended by the bisector of the incident and diffracted beam and the surface normal, d_ψ is the lattice spacing at a given ψ tilt and d_o is the unstressed lattice spacing.

Residual stresses are measured using either of two techniques. The first is the single exposure technique (SET), whereby a stress measurement is performed using only one beta β angle or tilt. This technique gives the user a very quick and efficient method to perform a stress measurement and is particularly suited for the need to take many measurements very quickly. The second is the multiple exposure technique (MET), whereby multiple beta tilts are used in the analysis.

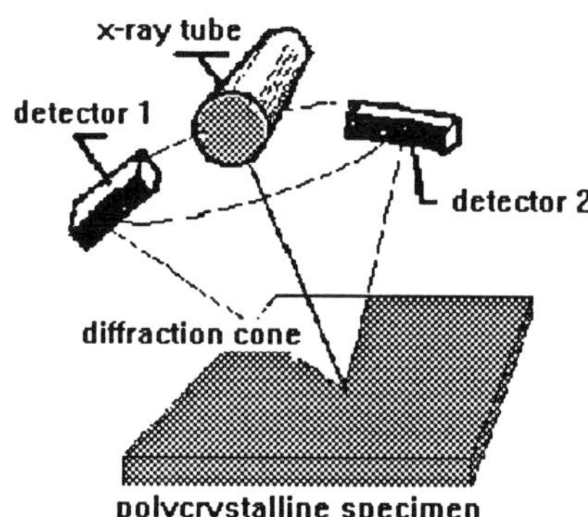

FIGURE 1 - HERE IS A DIFFRACTOMETER SETUP FOR STRESS MEASUREMENT WITH THE DIFFRACTION CONE OR DEBYE RING SHOWN. THE SOLID ANGLE OF THE RING IS DETERMINED BY THE BRAGG ANGLE OF THE DIFFRACTION PLANE.

The MET method is more revealing for materials for which the d vs. $\sin^2\psi$ relationship is not linear, as assumed in equation (3), but takes much longer than the SET (Klug and Alexander, 1974).

The stress measurements described in this paper were performed on iron alloys using both the SET and MET with Cr $K\alpha_1$ radiation and the (211) hkl plane. The experimental error for each stress measurement ranged from 2 to 3 ksi.

SURFACE PREPARATION

In many situations because of paint, loose rust, severe pitting or scale, some form of surface preparation was required. In the case of paint a solvent was used whereas in other cases a portable electropolisher was employed to prepare the surface for measurement. Electropolishing techniques are utilized because they enable the removal of material layers without cold working and introducing additional residual stresses.

MONORAIL BOX GIRDER

A number of steel box girders on a monorail were found to exhibit a camber inconsistent with design specifications. The proper camber was restored using a technique referred to as "heat straightening". It was the changes in residual stress resulting from this heat straightening procedure that were of great interest.

The heat straightening technique involves localized heating combined with an applied load, in this case hydraulic jacks. The applied localized heating allows a small region of the girder to plastically deform under moderate load. When the section cools a resultant change in girder shape is achieved.

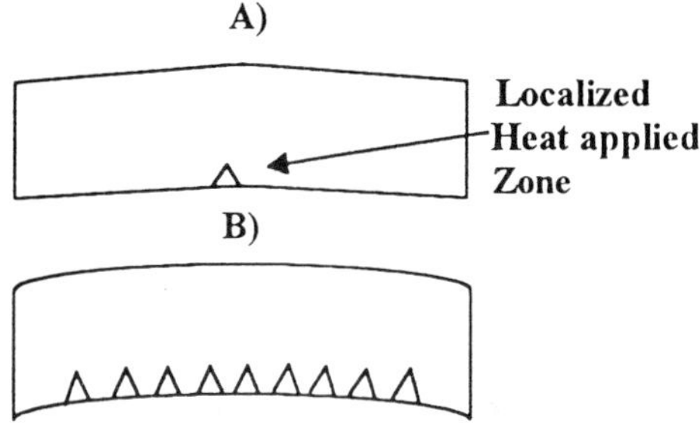

FIGURE 2 - A) THE EFFECT OF LOCALIZED HEAT STRAIGHTENING AT 1 LOCATION ON A GIRDER. B) THE FINAL RESULT WHEN SEVERAL POINTS ARE HEAT STRAIGHTENED.

The process is repeated at several locations along the length of the member until the desired camber is achieved. (see figure 2)

Residual stress profiles were collected across a typical heat straightened zone. The stress gradient and tensile residual stress field established can be seen in figure 3.

It was discovered that although a desirable camber was achieved, the resultant stress field that produced this camber may be detrimental to the fatigue life of the member.

It has been known for some time that tensile residual stresses in members that undergo tensile cyclic loading suffer from a reduction in fatigue life (Faires 1965). In addition finite element models can predict the applied loads on the system while fatigue models can a make determination as to whether short or long term fatigue damage will be more likely to occur.

These models can now be verified. Any changes in residual stress states during a member's service life can easily be monitored periodically using x-ray diffraction techniques. Subsequently, the effects of the fatigue cycling at the heat straightened locations can be characterized.

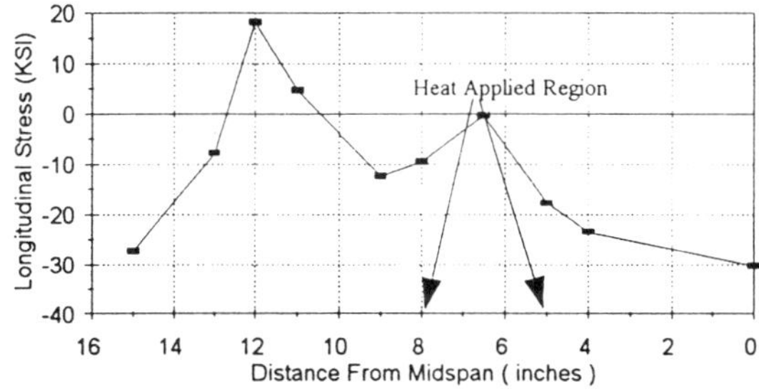

FIGURE 3 - RESIDUAL STRESS PROFILE ACROSS A HEAT STRAIGHTENED SECTION OF GIRDER WITH TENSILE RESIDUAL STRESSES.

FIGURE 4 - A TYPICAL HEAT STRAIGHTENED BOX GIRDER. HERE THE DIFFRACTOMETER IS SHOWN MEASURING STRESSES IN REGIONS OF LOCALIZED HEATING.

Another area of concern in the installation of monorail girders is the residual stresses due to welds. These are of particular interest because high stress gradients can also be created by the large amounts of differential heating and cooling inherent in the welding process. Typically, a tensile residual stress field is created in the heat affected zone as weld material cools and pulls on the parent material.

A stress gradient profile was performed across a weld splice where two monorail girders were welded together. Tensile residual stress maxima and a high stress gradient were detected. (see figure 5)

The tensile residual stresses in the HAZ of this weld splice were of course a concern since they pose a potential reduction in fatigue life.

As well, there is a potential for low cycle fatigue damage in both the heat straightened and welded areas since under maximum applied load, the yield strength of the material is exceeded at the tensile residual stress maxima.

In summary, stress fields in the monorail box girders were successfully characterized at numerous locations.

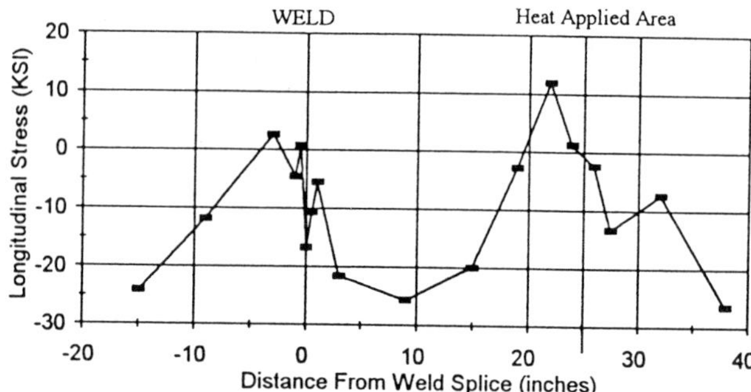

FIGURE 5 - THE EFFECT OF LOCALIZED HEAT STRAIGHTENING AND A WELD SPLICE HAZ.

BUILDING COLUMNS AND MEMBERS

Several stress measurements were performed on the main support columns of a multiple story building. A large amount of plastic deformation had taken place due to an impact absorbed by several support members.

The absolute (residual and applied) stresses were characterized on all four faces of each column. This was done so that an assessment of the damage could be quantified in terms of stress so that the potential for failure (in this case buckling) could be determined.

The stresses around the columns were mapped to determine if any unusually high stresses due to plastic deformation from the impact existed. An example of a map is shown in figure 6.

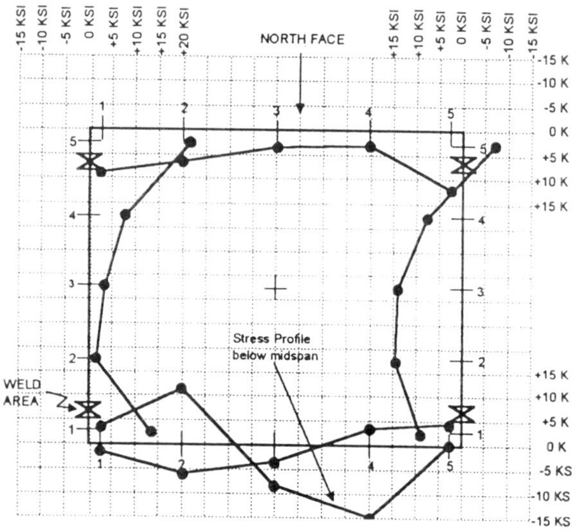

FIGURE 6 - STRESS MAP   COLLECTED AROUND A DAMAGED COLUMN.

FIGURE 7 - HERE A DIFFRACTOMETER IS BEING USED TO MAP STRESSES AROUND A DAMAGED COLUMN.

The results indicated that stress levels varied around the columns and that stresses in the areas of high plastic deformation were different than those in areas of less deformation. An accurate assessment of the stresses existing in these columns was performed.

SUSPENSION BRIDGE CABLES AND EYE BARS

Calibration of the elastic constants for bridge eye bar and cable material was performed in the lab using x-ray diffraction in conjunction with uniaxial tensile test fixtures.

An eye bar was placed in uniaxial tension with a known set of applied loads. Simultaneous axial stress measurements were performed using x-ray diffraction techniques. The applied stress was calculated for each incremental loading of the test specimen by dividing the applied load by the cross sectional area of the eye bar. The loading data were compared with the x-ray diffraction data. The slope of the measured x-ray axial stress vs. applied stress should be 1.0 if the experimental data were in agreement with theory. The slope of the experimental data was calculated to be 1.04 +/- 0.12 ksi. The resultant experimental data were thus in good agreement with theory. (see figure 8)

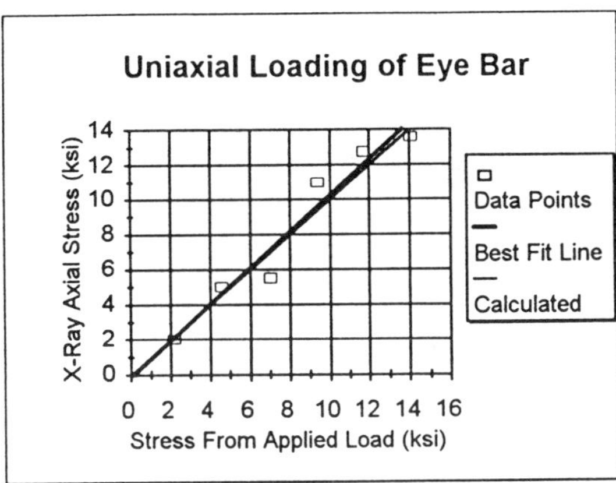

FIGURE 8 - CALIBRATION PLOT OF X-RAY AXIAL STRESS MEASURED VS. STRESS FROM APPLIED LOAD FOR AN EYE BAR.

The theoretical values of stress were calculated by dividing the applied load by the cross sectional area of the member.

The purpose for these trials was twofold. The first was to observe a response in the stress levels with the applied loads on bridge materials and the second was to determine the x-ray elastic constant to be used in the conversion of strains to stress values.

FIGURE 9 - X-RAY ELASTIC CONSTANT DETERMINATION ON AN EYE BAR.

FIGURE 10 - MEASURING STRESS ON A WIRE STRAND

A similar calibration procedure was applied to an individual wire in order to determine it's stress load characteristics. (see figure 10)

These lab experiments were completed prior to stress measurements performed in the field at the footings of a suspension bridge.

Since the bridge had been designed and built over 90 years previously, at a time when accurate modeling techniques did not exist, engineers essentially overdesigned to ensure structural integrity. Over time, due to minimal maintenance as well as ever increasing live load demands, the safety factor is diminished. X-ray diffraction was demonstrated to provide answers pertaining to the stress levels existing in the eye bars at the bridge footings.

The miniaturization of the x-ray diffraction equipment was invaluable in accessing the suspension cable footings since many steep staircases and service hatches would have made it extremely difficult if not impossible to bring in any larger equipment.

Several stress measurements were performed to determine the stress levels existing in both the main suspension cables and in the eye bars. The total stress on the eye bar measured at the footing was +24 ksi.

This result is in agreement with the predicted result calculated using a finite element model.

FIGURE 11 - NONDESTRUCTIVE ABSOLUTE STRAIN MEASUREMENT ON INDIVIDUAL STRANDS IN BRIDGE SUSPENSION CABLE, HERETOFORE CONSIDERED IMPOSSIBLE.

Diffraction measurements in this case were able to verify the model predictions, subsequently increasing the confidence in the predicted results.

Measurements were also performed on a cable strand with a measured stress of +20 ksi.

This kind of information had never before been obtained nondestructively with such a degree of accuracy and is vital in the proper care and maintenance of such a structure.

CONCLUSIONS

Accurate nondestructive stress measurements can be performed on steel structures via x-ray diffraction stress measurement techniques in the field. Absolute stress measurements were successfully completed on several types of steel structures, which would not have been feasible in the past.

Thus a practical tool now exists whereby engineers can, through nondestructive means, determine the actual stress state of any steel structure new or old and make informed decisions with a high level of confidence.

Furthermore, the nondestructive nature of the x-ray diffraction technique offers new approaches to the assessment of structures. For example, engineers can now take numerous measurements at monthly intervals at the same locations on a structure and use the data as a means of tracking the changes in the stress state of the structure. This data could eventually be used as a predictive tool, for use in optimizing the allocation of maintenance budgets

or in analyzing the state of a failure critical structural member. With x-ray diffraction it is now possible to accurately characterize localized stress gradients, for example, at or near welds, potentially important in fatigue life assessments.

In addition the engineer now has the means to test the accuracy and appropriateness of a given computer model or stress analysis technique by obtaining set of benchmark measurements through x-ray diffraction.

The speed and versatility of the latest generation x-ray diffraction equipment has positioned it presently as the most practical tool available today for the accurate nondestructive characterization of residual stress and absolute stress in structures in the field.

EQUIPMENT USED

All measurements were performed using a Proto XRD-1000[c] diffraction head powered by the Proto iXRD[c] portable x-ray diffraction unit.

REFERENCES

C.O. Ruud, 1982, "Nondestructive and Semidestructive Methods for Residual Stress Measurement", Residual Stress Effects in Fatigue, ASTM, Baltimore, pp. 3-5.

K.F. Dunker, B.G. Rabbat, March, 1993, "Why America's Bridges Are Crumbling", Scientific American, pp. 66-72.

M.E. Hilley et.al., 1971, "Residual Stress by X-Ray Diffraction - SAE J784a", Society of Automotive Engineers, Inc., Warrendale, PA.

J.A. Pineault, M.E. Brauss, 1994, "Insitu Measurements of Residual and Applied Stresses in Pressure Vessels and Pipelines Using X-ray Diffraction Techniques", PVP-Vol.276/NDE-Vol. 12, Determining Material Characterization: Residual Stress Integrity With NDE, ASME, New York, pp. 145-148.

V.M. Faires, 1965, "Design of Machine Elements", 4th Edition, The Macmillan Company, Collier-Macmillan Canada, Limited, Toronto, Ontario, pp. 100.

H.P. Klug, L.E. Alexander, 1974, "X-Ray Diffraction Procedures", 2nd Edition, Wiley-Interscience, U.S.A.

I.C. Noyan, J.B. Cohen, 1987, "Residual Stress Measurement by Diffraction and Interpretation", Springer-Verlag, New York.

APPLICATION OF THE L_{CR} ULTRASONIC TECHNIQUE FOR EVALUATION OF POST–WELD HEAT TREATMENT IN STEEL PLATES

Don E. Bray
Department of Mechanical Engineering
Texas A&M University
College Station, Texas

Paul G. Junghans
Interpretation, Research and Advanced Development
Atlas Wireline Services
Houston, Texas

ABSTRACT

Ultrasonic travel-times obtained with the L_{CR} wave are able to distinguish between stress relieved and non-stress relieved 13 mm (1/2 in.) thick, steel plates. Two 1.22 m (48 in.) square plates were patch welded in the center to create a residual stress field, and one of the plates was stress relieved. The L_{CR} ultrasonic technique sends a critically refracted longitudinal wave traveling beneath the plate surface, and the stresses in the plate affect the travel-times through the acoustoelastic relationship. The L_{CR} travel-time measurements not only distinguished between residual stress states in the plate, but also gave some information on their distribution and magnitude. Neither texture nor localized residual stresses affect the results. These findings demonstrate the potential usefulness of the technique for evaluating the state of post-weld heat treatment in structural steels.

INTRODUCTION

Residual stresses always have been a nemesis of sound engineering design. Intuitively, it is clear that compressive residual stresses countering tensile applied stresses can increase the load bearing capability of a structural item. Residual tensile stresses, on the other hand, can severely reduce this capability. The long-standing interest in residual stress has not prevented very common structural failures from unfavorable stress patterns. Many examples of residual stress related failures in structures are available in the technical references.(1,2)

Uncertainty in knowing the residual stress fields has led to significant over design in many engineering structures.(3) Wylde specifically describes the influence of residual stresses on designing welded structures.(4) The general conclusion is that in the absence of information about the residual stress, the safe assumption is that residual stresses may be as high as the yield strength of the material. Clearly, a knowledge of the residual stresses would be beneficial to engineering design. More specifically, knowing if a weld has been properly stress relieved would remove a large portion of the design uncertainty that now exists. The L_{CR} ultrasonic technique described here has demonstrated an ability to determine the stress relief state of welds in structural steel samples.

Virtually all of the physical phenomena used in NDE have been investigated for a capability to measure stresses in materials. While all have shown some success in particular applications, none has emerged as a fully accepted shop and field tool for the measurement of applied and residual stress in engineering structures. However, the acoustic techniques have demonstrated unique merit. A particular advantage of all of the acoustic techniques is that they rely on ultrasonic waves propagating through the material and, therefore, are capable of nondestructive stress measurements in a bulk region of the material.

The acoustic techniques are based on acoustoelastic effect which correlates the stress field in a material to changes in the ultrasonic wave speed. This linear relationship of stress on ultrasonic velocities is relatively small, on the order of a few hundreds of a percent(5); and yet, it has been used successfully for many years to measure bulk and near-surface stresses. In certain alloys, it has been suggested that this linear relationship persists even beyond the yielding point of the material.(6) Allen et al.(7) and Pao et al.(8) have compiled excellent overviews of the research that has been done on acoustoelasticity and ultrasonic stress measurement.

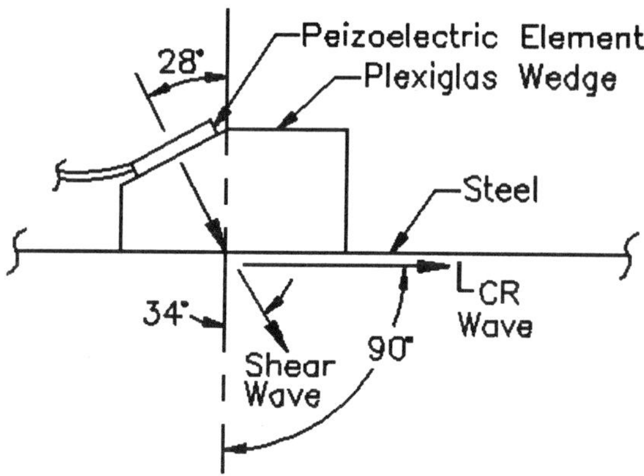

Fig. 1   Excitation of the L_{CR} wave in steel with an angle beam transducer.

Specific examples of acoustoelastic stress measurement are too numerous to reference but are almost exclusively restricted to the simple case of uniaxial stress fields.

CHARACTERISTICS OF THE L_{CR} ULTRASONIC STRESS MEASUREMENT TECHNIQUE

As research on the nondestructive measurement of stress has evolved over the years, it has become clear that the critically refracted longitudinal (L_{CR}) wave technique offers some distinct advantages over other methods. The L_{CR} wave may be generated with a critically refracted, angle beam transducer as shown in Figure 1 for a Plexiglas and steel combination. Since the L_{CR} wave is a bulk wave which travels just below the surface of the material, it is sensitive to a stress field in a finite thickness and not just at the surface. Additionally, stress gradients may be measured since the depth of penetration may be varied through the choice of different excitation frequencies.(9) Also important is that in comparison to the shear wave, the L_{CR} wave is the one most sensitive to stress and, yet, it is least sensitive to material texture.(10, 11 and 12)

The L_{CR} probe for the present experiments is a long device about 305 mm (12 in.) in length, and arranged in a pitch and tandem-catch fashion. The L_{CR} pulse is excited by the single transmitter and the travel-time is measured from successive arrivals at the two in-line receivers, as shown in Figure 2. Since the distance between the receivers is fixed, the acoustic velocity is directly proportional to the travel-time. The tandem-catch arrangement minimizes travel-time measurement errors that might be created by wave speed changes in the transducers or transmitter triggering uncertainty. Each transducer is comprised of a critically refracted Plexiglas wedge fitted with a 2.25 MHz, 25 mm (1 in.) square, air-backed piezoelectric element. Additional details on the angle beam L_{CR} probes and beam profiles are given by Junghans and Bray.(13)

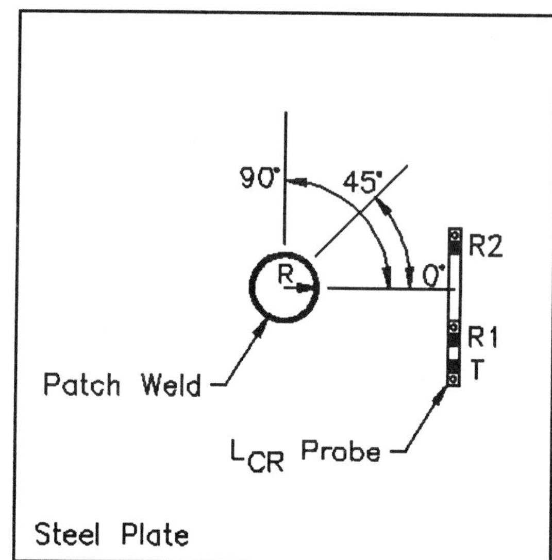

Fig. 2   Schematic of the L_{CR} probe and the patch welded plate showing the tangential orientation of the probe and the geometry of the test plan.

CONSIDERATION OF THE ULTRASONIC STRESS MEASUREMENT PARAMETERS

One problem associated with using the acoustoelastic relationship for stress measurement is that nominally identical materials may exhibit slight differences in wave speeds and acoustoelastic constants. The wave speed variations may originate from texture as well as from residual stresses. Differences in acoustoelastic constants most likely are caused by texture effects.(6) Where texture is relatively constant, however, the data indicate that stress differentials are measurable. Additionally, data collection and analysis schemes designed for specific applications may yield reasonable estimates of absolute stress levels.

There is considerable interest in nondestructively measuring the stress levels in structural members due to an applied force. To do this, however, the effects on the wave speed of the residual stress, texture, and temperature in the item must first be established.(14) For example, a structural steel member under active load yields a travel-time measurement containing the combined effects of applied force, residual stress, texture, and temperature. This is expressed by:

$$t = t^* + \Delta t_F + \Delta t_{RS} + \Delta t_{TX} + \Delta t_T \qquad (1)$$

where

t	- measured travel-time,
t^*	- travel-time for a homogeneous, isotropic, stress-free member at a standard temperature,
Δt_F	- travel-time effect of the applied force,

Δt_{RS} - travel-time effect of the residual stresses,

Δt_{TX} - travel-time effect of the material texture,

and

Δt_T - travel-time effect of the temperature difference at time of measurement from a standard temperature.

For ultrasonic measurement of the applied stress, the parameter of interest is Δt_F while the parameter which can be measured is t. The task is then to devise a scheme to either measure the effects of the remaining parameters, or to collectively deal with them as a group. The effects of temperature are well known.(15, 16) The possibility of separately identifying t^*, Δt_{RS} and Δt_{TX} is one which is appealing; yet, in practice this is very difficult. The question to be asked is whether or not the separation of every component is really necessary? In situations where either the applied or residual stresses dominate there may be no need to separately quantify them. Examples of this situation are welds, where the thermal induced stresses may be significantly larger than the load stresses, and structures where the undetected failure of one member can significantly increase the applied load on other members. In both of these circumstances the detection of high stress levels is important, regardless of whether the cause is residual or applied.

A similar approach to obtaining a reasonable measurement of applied stress in structural steel has been discussed by Williams.(17) He proposed obtaining travel-time data at several locations so that textural and residual stress variations will be averaged to a low level. Thus, the applied stress value will emerge as the dominant measured parameter. Bray and Leon-Salamanca indicate that this average technique might be useful in determining thermally induced stresses in railroad rail.(14)

WELD STRESS MEASUREMENT WITH THE L_{CR} ULTRASONIC TECHNIQUE

For stress measurement in welded plates, the change in travel-time due to applied force, Δt_F, may be redefined as Δt_{WS}, the travel-time change due to the weld induced stress. The manufacturing induced residual stress in the original plate remains Δt_{RS}. The travel-time, t_o, in a typical as-manufactured steel plate without weld stresses and at a reference temperature may be written as:

$$t_o = t^* + \Delta t_{RS} + \Delta t_{TX} \qquad (2)$$

Substituting Equation 2 into Equation 1 yields an applicable equation for the change in travel-time due to the weld induced stress in the plate at any temperature, T, to be:

$$\Delta t_{WS} = \Delta t_F = t - t_o - \Delta t_T \qquad (3)$$

where t, in this expression, is the experimental travel-time which would be obtained from the welded plate that is being evaluated. With a knowledge of the weld induced change in travel-time and the appropriate acoustoelastic constant, the stress change created by the weld at any other point on the plate may be calculated.

The relationship of measured L_{CR} wave change in travel-time to the corresponding change uniaxial stress is given by Egle and Bray(11) to be:

$$\Delta\sigma = \frac{E}{L\,t_o}\,(t - t_o - \Delta t_T) \qquad (4)$$

where

$\Delta\sigma$ - change in stress,

E - Young's modulus,

and

L - acoustoelastic constant for longitudinal waves propagating in the direction of the applied stress field.

While the assumption of an uniaxial stress field is not correct for the patch-weld circumstance of the present paper, the stress field for many butt-weld arrangements may be approximated by a uniaxial field.

The temperature effect on wave speed can be included, if the conditions require that this be done. For pearlitic steel the temperature effect(11) has been determined to be:

$$\frac{\Delta C}{\Delta T} = 0.55\,\frac{m}{s\text{-}^{\circ}C} \qquad (5)$$

For the probe used in these experiments the temperature effect on travel-time was calculated to be 0.0034 $\mu s/^{\circ}C$. With this value, Δt_T may be established for the conditions at the time the data are obtained. The range over which the temperature fluctuated during this experiment, $22.6^{\circ}C$ to $23.2^{\circ}C$, produced no significant effect.

EFFECTS OF TEXTURE ON L_{CR} STRESS MEASUREMENT IN WELDED STEEL PLATE

Changes in the material texture in the plate and near to weld seams are thought to be a serious obstacle to using ultrasonic techniques for stress measurement. This was not the case for two structural steel plates joined at the center by a longitudinal double vee groove weld, as reported by Leon-Salamanca and Bray.(18, 19) Data were obtained from the plates before and after heat treatment, using both the L_{CR} and neutron diffraction techniques. The effects of the stress relief were clearly shown by the L_{CR} data, but not by the neutron diffraction results which would be affected by the texture change. Since the heat treatment did not affect the texture, the results indicate that the initial travel-time profiles were caused by the internal stress of the weld and not by the texture.

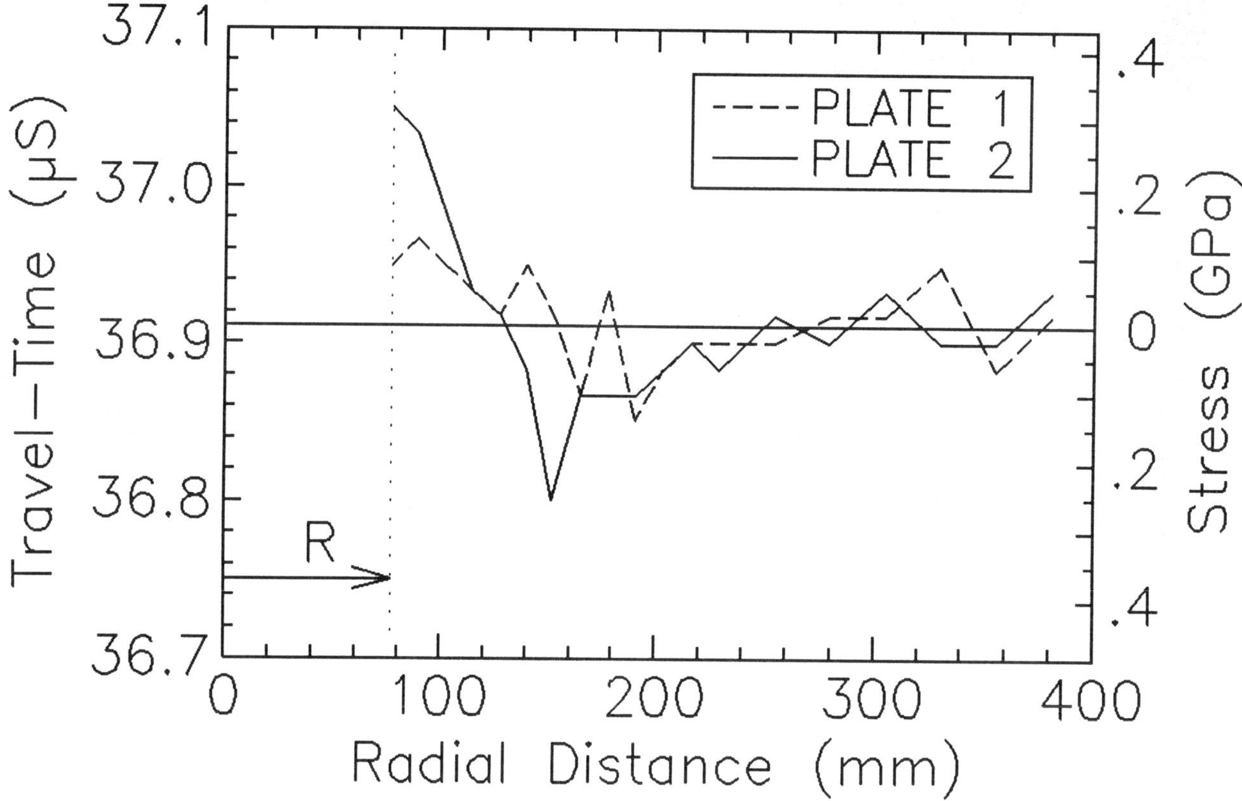

Fig. 3   Average differential travel-times for the L_{CR} waves travelling tangentially to a circular patch weld, R, as a function of the radial distance from the center for a stress releived plate (No. 1) and non-stress releived plate (No. 2).

APPLICATION OF THE L_{CR} TECHNIQUE ON STRUCTURAL STEEL PLATES

The present paper describes an additional evaluation of the L_{CR} technique for weld stress measurement using two patch-welded steel plates which were supplied by the DuPont Co. The test plates are made of 13 mm (1/2 in) thick, A516 Grade 60 steel. Each plate is 1.22 m (48 in) square and contains a 152 mm (6 in) diameter circular patch-weld in the center, as shown in Figure 2. One of the test plates had been stress relieved after it was patch-welded. The effects of the weld induced stress, localized residual stress and material texture on the velocity of the L_{CR} waves were examined in this experiment.

Theoretical analysis of patch-welds predicts that the largest variation in the stress distribution will occur in the tangential component as a function of the radial distance.(20) Thus, the tangential travel-time measurements taken in these experiments along several radial directions were the ones most likely to detect the stress gradient created by the patch-weld. The additional effect of texture induced anisotropy was also investigated by varying the direction of propagation of the L_{CR} waves

in the plates.

A full factorial design was used in this experiment. A diagram of the test plan that was used for the travel-time measurements is shown in Figure 2. The L_{CR} wave probe was aligned tangential to the weld bead. Travel-time measurements were taken along 3 radial directions (0°, 45°, and 90°) on each plate. Along each radial line, data were obtained at 19 distances. Near to the patch, where the stress effects would be greatest, data were obtained at radial distances of 76 mm to 230 mm (3 in. to 9 in.) in increments of 13 mm (1/2 in.). From 230 mm to 380 mm (9 to 15 in.) data were taken at 25 mm (1 in.) intervals. Temperature measurements were taken at each location during the differential travel-time measurement procedure. This test plan produced 114 observations, 57 per plate. Randomization of the data taking process was used to eliminate bias and to insure random distribution of the experimental variance.

The received signals were digitized with a computer system called **PCDAS**, Personal Computer Data Acquisition System, which allowed 16 wave forms to be averaged for each travel-time measurement. This system uses an analog to digital converter board capable of real-time sampling rates of 20 MHz. This is notable since the accuracy of the travel-time measurements is a direct function of the sampling rate of the data acquisition system. Overall accuracy in measuring the arrival times was felt to be at least 0.025 μs.

The ultrasonic travel-times obtained with the L_{CR} wave were used to distinguish between the stress relieved and non-stress relieved steel plates. Figure 3 is a plot of the average differential travel-times for each plate as a function of the radial distance from the patch weld. Each data point is the mean travel-time of the three different angular samples at the same radial distance. Plate 1 was determined to be the stress relieved plate, while plate 2 was in the as welded condition. A pattern in the data is apparent for plate 2. The data points farthest from the weld are relatively flat. Approaching the weld, there is a minimum travel-time at 152 mm (6.5 in) followed by a steep rise to a peak at the weld radius, 76 mm (3 in.). While there is scatter in the data for both plates, no distinctive minimum or maximum can be found for the stress-relieved plate 1.

The data pattern presented in Figure 3 for plate 2 corresponds well to the theoretical predictions for the tangential weld stress distribution that would be created by a circular patch weld in a larger plate(1) A region of high tangential, tensile stresses exist at the weld due to circumferential shrinkage of the weld metal. These stresses may be as high as the yield strength of the material. Farther out from the patch-weld, a region of lower tangential, compressive stresses exist due to radial shrinkage of the patch-weld. Beyond this compressive region, a stress free area exists where the material is not affected by the patch-weld. This region may be used to determine the zero-stress ultrasonic velocity for the material. All of these regions may be distinguished in the experimentally determined tangential travel-time distribution of the non-stress relieved Plate 2.

ESTIMATION OF THE WELD INDUCED STRESSES

By assuming that the mean travel-time data in the region from 230 mm to 380 mm (9 to 15 in) represent a zero-stress travel-time base, the relative change in stress at any point on the plate may be calculated from travel-time measurements at that point by using Equation 4. The mean zero-stress travel-times for plates 1 and 2 were 36.9119 μs and 36.9095 μs, respectfully; thus, yielding an average zero-stress travel-time, t_o, of 36.911 μs with a standard deviation of 0.056 μs. The greatest difference between the zero-stress travel-time and a measured travel-time was at the weld in plate 2, $t - t_o = 0.139$ μs. The range over which the temperature fluctuated during this experiment, 22.6°C to 23.2°C, produced no significant change in L_{CR} wave speed; thus, effect of temperature, Δt_T, is assumed to be zero.

From Egle and Bray(11), the appropriate acoustoelastic constant for pearlitic steel is 2.45. Young's modulus for the steel is 207 GPa (30 X 10^6 psi). For approximation

purposes, it will be assumed the present data from the patch welded plates is representative of data from a uniaxial stress field. Thus, the stress difference calculated using Equation 4 for the maximum difference obtained for the present experiments is:

$$\Delta s = \frac{(207 \times 10^9)(0.139)}{2.45(36.911)} = 318 \text{ MPa (46 ksi)}$$

This value exceeds the minimum yield stress of 262 MPa (38 ksi) for the structural steel used for the patch welded plates. The calculation, however, is based on the assumption of a uniaxial stress state. In reality, a biaxial stress state exists at the weld seam of a circular patch weld. The stress scale shown in Figures 3 is only an approximation of the tangential stress in the weld that was derived from the assumption of a uniaxial state of stress.

STATISTICAL ANALYSIS OF THE DATA

Data were analyzed using an analysis of variance (ANOVA) and a least significant difference T test; procedures widely used for identifying the significant independent variables within data sets and comparing pairs of means with equal variances.(21, 22) The analysis of variance procedure partitions the total variation in a measured response, such as travel-time, into components which can be attributed to recognizable sources of variation.(23) Significant factors within data sets are indicated by a F value statistic, which is the ratio of the variation attributed by the given treatment (factor) to the variation attributed by random experimental error.

A significance level of one percent ($a = .01$) was selected for this analysis. The significance level, a, indicates the probability that a variable will be identified as significant when, in truth, it is not. For a given significance level, the number of treatments (k) and the total number of data points (N) establishes a minimum critical F value, which may be obtain from statistical tables. The statistical computer program used for this analysis went one step further. It calculated the actual significance probability, Pr, corresponding to each F value. If this calculated significance level, Pr, is less than or equal to the selected level, a, the null hypothesis, that all of the samples are the same, may be rejected; and thus, the conclusion is that the treatment had a significant effect on the measured response. Otherwise, the samples are assumed to be the same.

The data for the present investigation were first analyzed to detect any variation with radial angle in the zero-stress region. Tables 1 and 2 show the ANOVA results for data obtained in the unstressed region at the three angles in plates 1 and 2, respectively. Since, in both cases, the significance probability, Pr>F, is greater than the selected $a = 0.01$, there is a 99% confidence that there are no significant differences by angle in the

Table 1 Analysis of variance (ANOVA) of travel-times by angle in unstressed region of plate 1 ($\alpha = 0.01$).

Source	DF	SS	MS	F value	Pr>F
Angle	2	0.0267	0.0133	3.54	0.0506
Error	18	0.0678	0.0038		
N	20	0.0945			

Table 2 Analysis of variance (ANOVA) of travel-times by angle in unstressed region of plate 2 ($\alpha = 0.01$).

Source	DF	SS	MS	F value	Pr>F
Angle	2	0.0074	0.0037	2.58	0.1032
Error	18	0.0257	0.0014		
N	20	0.0331			

Table 4 Analysis of variance (ANOVA) of travel-times by radius for plate 1 ($\alpha = 0.01$).

Source	DF	SS	MS	F value	Pr>F
Radius	18	0.0524	0.0029	0.60	0.8739
Error	38	0.1833	0.0048		
N	56	0.2358			

Table 5 Analysis of variance (ANOVA) of travel-times by radius for plate 2 ($\alpha = 0.01$)

Source	DF	SS	MS	F value	Pr>F
Radius	18	0.1860	0.0103	3.10	0.0016
Error	38	0.1267	0.0033		
N	56	0.3126			

data obtained in the unstressed regions of the plates. Further, Table 3 shows the ANOVA results comparing the two plates using the travel-time data in the unstressed regions. Again, the statistics indicates that the travel-times in the unstressed regions of the two plates are not significantly different. To determine if significant differences exist in any region of plates 1 and 2, an analysis of variance by radial distance was run in both plates. The results are shown in Tables 4 and 5. For plate 1 (Table 4), the Pr>F is greater than the selected α, indicating that there are no significant differences in travel-time by radius in the plate. For plate 2 (Table 5), however, the results indicate that significant differences in the travel-time do occur with radius since Pr>f is less than $\alpha = 0.01$.

Unfortunately, the ANOVA procedure does not tell us at which of the radial distances in plate 2 may the travel-times be regarded as significantly different from the others. Therefore, a least significant difference T test was used to identifying the differences in data set. A significance level of one percent, $\alpha = 0.01$, was again

selected. The results of the T test are reported in groupings of means signified by different letters, starting with A. Where significantly different data groupings are found, these are identified with different letters. Means within a group may also be different; but, we cannot claim this with a 99% confidence.

Table 6 shows the T test groupings of the means for all of the data collected from plate 2. A graphical representation of these groupings are shown in Figure 4. The smallest group, A, contains the four points nearest to the weld having the longest travel-times, as well as two in the outer regions. Group C contains all of the points in the assumed zero-stress area of the plate. Group D includes points with the shortest travel-times and particularly the minimum at 152 mm. These groups corresponds well to the theoretical predictions for the tangential stress distribution in a patch-welded plate.

Several important observations may be made from the data analysis. First, in the region from 230 mm to 380 mm, the zero-stress area, the manufacturing induced residual stresses and texture in the plate should dominate over weld induced stresses. The analysis of data in this region showed no significant differences in travel-times by angle or plate. These two findings lead to the conclusion that the average L_{CR} data are not significantly affected by texture and localized residual stress.

The significantly different data found for the non-stress-relieved plate in the region near to the weld (region A) indicates that the L_{CR} technique is capable of distinguishing stress relieved welded steel plates from those that have not been stress relieved. While the increase is much larger for plate 2, both plates show some increase in the travel-time at the weld and in the

Table 3 Analysis of variance (ANOVA) of travel-times by plate in the unstressed region ($\alpha = 0.01$).

Source	DF	SS	MS	F value	Pr>F
Plate	1	0.00006	0.00006	0.02	0.8920
Error	40	0.1276	0.0032		
N	41	0.1277			

Table 6 Groupings by radius from the least significant difference T test for the mean travel-times in plate 2 ($a = 0.01$).

Grouping				Mean (μs)	N	Radius (mm (in))
	A			37.050	3	76.2 (3.0)
B	A			37.033	3	88.9 (3.5)
B		A	C	36.983	3	101.6 (4.0)
B	D	A	C	36.933	3	114.3 (4.5)
B	D	A	C	36.933	3	381.0 (15.0)
B	D	A	C	36.933	3	304.8 (12.0)
B	D	A	C	36.917	3	127.0 (5.0)
B	D	A	C	36.917	3	254.0 (10.0)
B	D		C	36.900	3	279.4 (11.0)
B	D		C	36.900	3	355.6 (14.0)
B	D		C	36.900	3	330.2 (13.0)
B	D		C	36.900	3	215.9 (8.5)
B	D		C	36.883	3	139.7 (5.5)
B	D		C	36.883	3	203.2 (8.0)
B	D		C	36.883	3	228.6 (9.0)
	D		C	36.867	3	177.8 (7.0)
	D		C	36.867	3	165.1 (6.5)
	D		C	36.867	3	190.5 (7.5)
	D			36.800	3	152.4 (6.0)

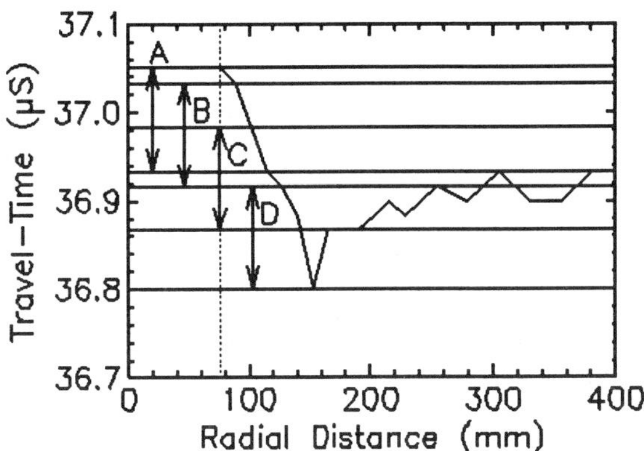

Fig. 4 T test groupings by radius of the average differential travel-times for the L_{CR} waves propagating in non-stress relieved plate 2.

small area near to it. These results clearly indicate that plate 1 was stress relieved. They may also indicate that this stress relief was not complete.

The scatter which appears in the data could be derived from several sources, including local residual stresses and texture as previously discussed. Most likely, much of this scatter is a result of random experimental error in the travel-time picks and variations in the coupling of the ultrasonic probe to the plates. A shorter probe configuration would, most likely, reduce these coupling induced variations.

USING THE L_{CR} TECHNIQUE FOR WELD STRESS MEASUREMENT

The previously cited results on the welded plates suggest an immediately available procedure that could determine the presence or absence of high stresses in a weld and in the surrounding area.(19) The procedure would be to obtain a travel-time profile across the region being investigated, starting at an assumed stress free reference point that is away from the weld affected zone. Probe orientation is held constant with respect to the plate's rolling direction to control the effect of the texture on the ultrasonic velocity measurement. A characteristic pattern in the travel-time profile would indicate that the part has not been stress relieved. The absence of a significant travel-time pattern would indicate a stress relieved area. Without question, further research is needed to establish full confidence in the ability of the technique to measure weld stresses. Nonetheless, the results obtained thus far show that the L_{CR} method can be used to detect weld induced stresses.

A recent report by Kobayashi et al.(24) demonstrates the usefulness of the ultrasonic acoustoelastic techniques in stress measurement. Fatigue specimens containing a weld seam were removed from a large, welded steel pipe and machined to dimensions of 100 mm (3.94 in.) square and 7 mm (0.28 in.) thick. Prior to fatigue testing, the stress field in each specimen was investigated using the shear wave acoustic birefringence acoustoelastic technique. This technique uses shear waves propagated through the thickness of the sample. The stress field is evaluated at several locations by the velocity differences obtained in shear waves with particle motion alternately parallel and perpendicular to the expected principal stress direction. Peak longitudinal stresses were found at the weld seam, as expected, with lower values away from the seam. The ultrasonic measurements were compared with strain-gauge data and a good correlation was found. Peak tensile stress values above 100 MPa (14.5 Ksi) were found for all of the samples. Fatigue crack growth predictions based on the measured values of residual stress agreed well with the experimental fatigue results. This study demonstrates the value of ultrasonic techniques in predicting fatigue life in welded specimens.

The shear wave acoustic birefringence acoustoelastic technique, described above, uses the travel-time change in multiple echoes propagating across the thickness of the sample. Clearly, that technique is limited to fully isotropic material having highly parallel sides. Moreover, the stresses measured will be the average through the thickness, and not indicative of any gradient. Since the L_{CR} technique is less sensitive to texture, most sensitive to stress, and capable of indicating stress gradients, it offers superior advantages in weld stress measurement over the acoustic birefringence acoustoelastic method. Further more, the L_{CR} technique does not impose any strict geometric limitations on the test specimens.

OTHER APPLICATIONS OF THE L_{CR} TECHNIQUE FOR STRESS MEASUREMENT

The L_{CR} technique has shown an ability to indicate stress conditions in a variety of materials. The earlier studies applying the technique to stress measurement in railroad rail(11, 14, 15 and 16) were continued by Polish investigators.(25, 26) They report success in monitoring manufacturing induced stresses as well as thermal induced stress changes in rail installed in the track. Also, the technique has demonstrated an ability to measure stress in castings made from ductile cast iron.(27, 28) In another application, the L_{CR} technique was used to determine the stress distribution in a shrunk fit retaining ring installed in a turbine rotor at an electric generating station.(29) As interest in the L_{CR} technique expands, further applications will surely develop.

CONCLUSIONS

The results show that the L_{CR} stress measurement technique can quantitatively distinguish between stress relieved and non-stress-relieved plates. Furthermore, it appears that the degree to which a plate or weld is stress relieved may also be indicated. Once the zero-stress travel-time is determined from a stress free region of a plate, the relative stress level at any other point on the plate may be calculated from travel-time measurements at that point and the appropriate acoustoelastic constant. The patch-welded plate's stress distribution that was approximated from the experimentally measured travel-times have been confirmed by other experimental methods as well as theoretical analysis.

The overall precision of stress measurement with the L_{CR} technique is dependent on the accuracy of the system used to obtain travel-time data. Where the system used for the present work had an accuracy of 0.025 μs, systems presently available are capable of measurements within 0.001 μs. Further, a new smaller probe design as used in the ductile cast iron studies(23, 24) will improve the precision of the overall system by reducing the scatter caused by poor coupling due to plate warpage.

Uncertainty which exists in using the L_{CR} technique generally rests in unanticipated variations in the parameters affecting the travel-times, namely the acoustoelastic constant, the presence of any external forces, and material texture. Field experience in using this technique has shown that variations are to be expected. Laboratory experimentation will continue to quantify the acousto-elastic constants for other materials, to develop more robust data acquisition methodologies, as well as to improve accuracy and precision of the travel-time measurement systems.

ACKNOWLEDGMENTS

Portions of this material were reported in a paper presented at the 1989 Pressure Vessels and Piping Conference of the American Society of Mechanical Engineers in Honolulu, Hawaii, July 1989. Partial support for this work was furnished by the Dupont Company and the Department of Mechanical Engineering of Texas A&M University. This paper has been submitted for publication in the Journal of NDT&E Int'l.

REFERENCES

1. Masubuchi, K., _Analysis of Welded Structures_, First Edition, Pergamon Press, New York (1980).

2. Wulpi, D. J., _Understanding How Metals Fail_, American Society for Metals, Metals Park, OH (1985).

3. Shigley, J. E., and Mischke, C. R., _Mechanical Engineering Design_, Fifth Edition, McGraw-Hill, New York (1989).

4. Wylde, J. G., "The Influence of Residual Stresses on the Fatigue Design of Welded Steel Structures," _Residual Stress in Design, Process and Materials Selection_, W. B. Young, Ed., American Society for Metals, Metals Park, OH (1987) pp. 85 - 95.

5. Schnieder, E., and Gobbels, K., "Nondestructive Evaluation of Residual Stress States Using Ultrasonic Techniques," _Residual Stresses_, Proceedings of the European Conference on Residual Stresses, 1983, E. Macherauch and V. Hauk, Eds., Informationsgesellschaft, Verlag, Germany (1986) pp. 247 - 261.

6. Thompson, R. B., Smith, J. F., and Lee, S. S., "Effects of Microstructure on the Acoustoelastic Measurements of Stress," _Nondestructive Evaluation: Application to Materials Processing_, Buck, O., and Wolf, S., Eds., The American Society for Metals, Metals Park, OH (1984) pp. 137 - 145.

7. Allen, D.R., Cooper, W.H.B., adn Silk, M.G., "The Use of Utrasonics to Measure Residual Stress," _Research Techniques in Nondestructive Testing, VI_, R.S. Sharpe, Ed., London (1982) pp. 151 - 209.

8. Pao, Y.H., Sachse, W., and Fukuoka, H., "Acousto-elasticity and Utrasonic Measurements of Residual Stresses," _Physical Acoustics_, W.P. Mason and R.N. Thurston, Eds., Academic Press, New York, NY (1984) pp. 61-143.

9. Leon-Salamanca, T., and Reinhart, E., "Surface Residual Stress Analysis of Metals and Alloys," Final Report, Department of the Navy, Naval Air Systems Command, Washington, D. C., Reinhart & Associates, Austin, Texas (1992).

10. Bray, D. E., and Stanley, R. K., _Nondestructive Evaluation Engineering_, McGraw-Hill, New York (1989).

11. Egle, D.M., and Bray, D.E. "Measurement of Acoustoelastic and Third-Order Elastic Constants for Rail Steel," _Jour. Acoustical Soc. Amer._, _60_, No. 3 (1976) pp. 741-744.

12. Bray, D. E., and Egle, D. M., "Ultrasonic Studies of Anisotropy in Cold-Worked Layer of Used Railroad Rail," _Metal Science_, _15_ (1981) pp. 574 - 582.

13. Junghans, P. G., and Bray, D. E., "Beam Characteristics of High Angle Longitudinal Wave Probes," NDE: Applications, Advanced Methods, and Codes and Standards, R. N. Pangborn et al., Eds., PVP-Vol. 216, NDE Vol. 9, Proceedings of the 1991 Pressure Vessels and Piping Conference, San Diego, CA, June 23-27, 1991, The American Society of Mechanical Engineers, New York, NY, (1991) pp. 39-44.

14. Bray, D.E., and Leon-Salamanca, T., "Zero-Force Travel-Time Parameters for Ultrasonic Head-Waves in Railroad-Rail" Materials Evaluation, *43*, No. 7 (1985) pp. 854-858, 863.

15. Egle, D.M., and Bray, D.E., "Nondestructive Measurement of Longitudinal Rail Stresses: Application of the Acoustoelastic Effect to Rail Stress Measurement" Federal Railroad Administration, Washington, DC, Report No. FRA/ORD-77/341, PB-281164, NTIS, Springfield, VA (1978).

16. Egle, D.M., and Bray, D. E., "Application of the Acousto-Elastic Effect to Rail Stress Measurement," Materials Evaluation. *37*, No. 4 (1979) pp. 41 - 46, 55.

17. Williams, H.D., Armstrong, D., and Robins, R.H., "Application of Ultrasonic Stress Measurements to Problems in the Electric Supply Industry," Jour. of Testing and Evaluation, *10,* No. 5 (1982) pp. 217-222

18. Leon-Salamanca, T., and Bray, D. E., "Ultrasonic Measurement of Residual Stress in Steels using Critically Refracted Longitudinal Waves (L_{CR})," Proceedings 1990 SEM Spring Conference on Experimental Mechanics, Albuquerque, NM, June 4-6, 1990, pp. 271-278.

19. Bray, D. E., Leon-Salamanca, T. and Junghans, P., "Application of the L_{CR} Ultrasonic Technique for Evaluation of Post-Weld Heat Treatment in Steel Plates," Nondestructive Evaluation NDE Planning and Application Proceedings 1989 ASME Pressure Vessels and Piping Conference, Honolulu, Hawaii, July 23-27, 1989, R. D. Streit, Ed., pp. 191-197.

20. Fukuoka, H., et al., "Acoustoelastic Stress Analysis of Residual Stress in a Patch-Welded Disk," Experimental Mechanics, *18*, No. 7 (1978) pp. 277-280.

21. Montgomery, D. C., Design and Analysis of Experiments, Third Ed., John Wiley & Sons, New York, NY (1991).

22. Hough, C. L., "The Effect of Back Rake Angle on the Performance of Small-Diameter Polcrystallline Diamond Rock Bits: ANOVA Tests," Journal of Energy Resources Technology, *108*, Dec. (1986) pp 305-309.

23. Milton, J.S., and Arnold, J.C., Probability and Statistics in the Engineering and Computing Sciences, First Edition, McGraw-Hill, New York (1986).

24. Kobayashi, H., Arai, Y., Ohsawa, Y., Nakamura, H. and Todoroki, A., "Nondestructive Measurement of Welding Residual Stresses by Acousto-Elastic Technique and Prediction of Fatigue Crack Growth," Jour. Pressure Vessel Technology, *114*, Nov. (1992) pp. 417-421.

25. Brokowski, A. and Deputat, J., "Ultrasonic Measurements of Residual Stresses in Rails," Vol. One, Proceedings 11th World Conference on Nondestructive Testing, American Society for Nondestructive Testing, Columbus, OH (1985) pp. 592-598.

26. Szelazek, J., "Ultrasonic Measurement of Thermal Stresses in Continuously Welded Rail," NDT & E Intl., *25,* No. 2 (1992) pp. 77 - 85.

27. Srinivasan, M. N., Bray, D. E., Junghans, P., and Alagarsamy, A., "Critically Refracted Longitudinal (L_{CR}) Wave Technique: A New Tool for Measurement of Residual Stresses in Castings," AFS (American Foundary Society) Transactions, 91-157, pp. 265-267.

28. Srinivasan, M. N., Chundu, S. N., Bray, D. E., and Alagarsamy, A., "Ultrasonic Technique for Residual Stress Measurement in Ductile Iron Continuous Cast Round Bars," Journal of Testing and Evaluation, *20*, No. 5 (1992) pp. 331 - 335.

29. Leon-Salamanca, T., "An Ultrasonic Method for Stress Measurement in Structures," Meeting EEI Metallurgy and Piping Task Force, Plaza Hotel, San Antonio, TX (1988).

A NONDESTRUCTIVE TEST CONTRIBUTION
TO THE SAFETY OF OFFSHORE CRANES

G. Hands
MEAGH
Measuring Techniques
Nieuwegein, The Netherlands

P. Bourgeors-Jacquet
SKF RKS SA
Avallon, France

INTRODUCTION

Non-Destructive Testing (NDT) plays a vital role in safety as well as in economics of manufacturing. This paper shows an example of how NDT has been successfully used offshore in Europe for a number of years, inspecting slewing bearings.

Slewing bearings are the bearings that support amongst other things cranes and earth-moving equipment. These bearings carry a very heavy slow moving unbalanced axial-load. Normal bearing design techniques are not used for these bearing types; slewing bearings usually have multiple row rolling elements.

A typical design of a slewing bearing is shown in Figure 1.

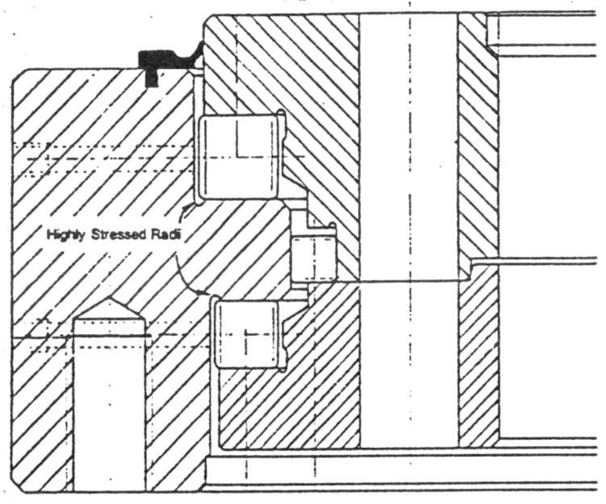

Figure 1 Typical design of a slewing bearing

Due to the extreme environment and the very heavy and unbalanced loading supported by the bearings, particularly on oil exploration and production platforms in the North Sea, some bearings have failed catastrophically. Such failures can have fatal consequences for the crane driver, apart from leading to unplanned unavailability of cranes, and very high extra costs, whilst replacement bearings are manufactured (this often taking several months).

In order to minimize such dangers, authorities have stipulated that slewing bearings for cranes employed in the North Sea are to be periodically inspected. This normally entails removing the bearing from the crane and replacing it with another bearing, whilst the first bearing is shipped onshore, dismantled and non-destructively inspected. The costs involved in this exercise are very high, as the crane is out of use for about a week and a team of men is required to inspect the bearing. In addition, proof lifts must be performed before the crane can be used again.

The slewing bearings are manufactured by a different process to those of the competitors, employing metallurgical techniques that ensure that any crack propagation is very slow. Bearings also have an additional safety feature for North Sea applications, that ensures that even with a sudden and total failure of the bearing, the crane will not fall off its mountings. In addition, SKF has developed and patented a design change on standard 3-row roller slewing bearings that allows an ultrasonic test to inspect the critical areas of the slewing bearing, without any requirement to dismantle the bearing from the crane.

In practice, such a bearing can be inspected in a few hours, often "fitting in" to periods when the crane is not in use. This leads to the possibility of more frequent monitoring of the bearing that the five years presently stipulated.

Catastrophic failure of slewing bearings is fortunately, not the failure mode commonly encountered in such industrial environments.

The normal life-restricting failure mode of slewing bearings is spalling of the raceways. This is the rolling contact surface of the bearing breaking-up under the very heavy loads imposed upon it.

This is normally detected either by analysis of grease coming out of the bearing, looking for metal particles, or the life of the bearing can be measured by the axial displacement of the crane as the raceway disintegrates, (in some cases, up to 5 mm displacement is permitted).

The developed ultrasonic testing technique, apart from detecting and monitoring cracks in the stressed areas of the bearing, is also able to monitor the initiation and growth of any spalling on the main raceways of the bearings, thus enabling better planning on maintenance and replacements.

This paper describes the development of the ultrasonic testing system.

TECHNIQUE DEVELOPMENT

Experiments to evaluate non-destructively the integrity of the highly-stressed radii of the slewing bearings were carried out initially in the laboratory on cracked bearing sections. These cracks were natural cracks and spark-eroded artificial defects, located in the highly stressed radius of the slewing bearing. This area is radiussed (or ground into a special shape, see Figure 1) to reduce the stresses, as stress analysis has shown this area to be the most likely to fail under extreme conditions of misuse.

Eddy current techniques were not very well suited to the application, because all the defects were in a difficult-to-access position, and furthermore, the defect orientation was not ideal for detection.

Several experiments with acoustic emission were performed, but due to the very slow rotation and heavy loading of such bearings in use, other emissions precluded detection of crack initiation and propagation with this technique.

Experiments with ultrasonics showed that the optimal detection technique was to use surface waves, propagated radially along the raceway and around the radius. This however was not considered to be a practical solution, as the raceway would normally be covered in grease, and would be regularly over-rolled by heavily-loaded rollers. A second ultrasonic technique, employing transverse waves propagating through the material to irradiate the radius at a suitable angle, proved to be more useful for the application, although difficult to implement and somewhat less sensitive than the surface wave technique.

To continue with the development, based on the transverse wave technique that had shown some success at detecting cracks

in the laboratory environment, some initial experiments were carried out in the factory on a full-size bearing using a specially manufactured longitudinal wave probe and a specially manufactured transverse wave probe. The probes were inserted through additionally-machined holes in a standard slewing bearing. See Figure 2.

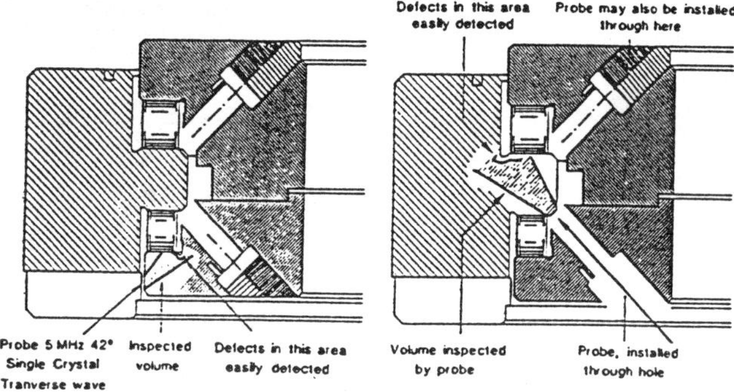

Figure 2 Ultrasonic probe positions for a standard slewing bearing

The longitudinal wave probe was a twin crystal type, 10 mm diameter, 5 MHz, and the transverse wave probe was a single crystal 10 mm diameter, 4 MHz probe, manufactured regarding probe angle, housing material and shape, etc. Stainless-steel wear areas on the probes ensure that these will have as long a life as possible.

To achieve adequate and reproducible coupling conditions, the longitudinal wave ultrasonic probe was manufactured with an oil fee tube running the entire length of the probe, to apply a coupling fluid (oil) immediately beneath the contact fact of the probe. This did not work well, as the diameter restrictions imposed on the permissible extra holes in the bearing meant that insufficient acoustic barriers were obtainable between the probe and the oil feed tube, giving large spurious signals from crosstalk between the send and the receive side of the probe. Smaller active elements within the probe were initially tried in order to overcome this, but the reduced sensitivity precluded the use of this type of probe. To overcome this difficulty, a new probe without couplant-feed was manufactured, and additional greasing holes were manufactured into the slewing bearing to allow grease to be injected into the required location immediately before testing.

With the probes located in one section of the bearing, and impinging on the other section of the bearing, rotation of the bearing will ensure that the probe is reproducibly and accurately scanned around the entire circumference. Small springs to apply a constant loading to the probe help to remove inaccuracies caused by probe pressure.

When testing is not being performed, the probes are withdrawn from the slewing bearing, to avoid unnecessary wear.

EXPERIMENTS

The experimental slewing bearing had an outside diameter of 1.73 meters, and was of the three-row rolling element type, with two rows of rolling elements carrying the axial load, and one smaller row of rolling elements carrying the radial load.

Crack initiation defects were manufactured into the highly stressed radii of the slewing bearing (using a high speed 0.15 mm thick cutting wheel). The purpose of these crack initiation defects was two-fold. Firstly, to act as reference targets for ultrasonic detection, and secondly, to act as crack initiators that would propagate during the trials with the bearing under load, otherwise the defects would be unlikely (or take an extreme test-time) to initiate or propagate, considering the normal quality criteria. The location of these defects was such that they were in the most highly loaded areas of the slewing bearing.

The defects were located in the three most highly stressed radii (numbered 1, 2 and 3), and the defects were identified as 11 and 12 on radius 1, 21 and 22 on radius 2, and 3 on radius 3.

The slewing bearing was assembled, and trials with a conventional portable ultrasonic flaw detector showed that the crack initiation defects were detectable by a skilled ultrasonic operator using manual evaluation.

Defect amplitudes (relative to a calibration reflector) and locations were manually recorded for comparison with future inspections.

The slewing bearing was mounted on a test installation at the factory in France, and then loaded to 60 tons load at a 2 meter radius from the rotation center, and rotated under load, reversing the rotation direction every day. Frequent manual ultrasonic tests were carried out on this bearing through its practically usable life (i.e., up to extensive spalling), but these failed to show any clear crack growth or propagation from the introduced defects. The ultrasonic evaluation here was recorded manually, by setting the instrument sensitivity to a known level and recording the amplitudes of the ultrasonic signals as interpreted on the screen.

The spalling on the raceways was apparent from the results of the ultrasonic tests, however, and the growth of the spalling was monitored throughout the life of the bearing, until 12825 revolutions under load had been performed.

At this stage, the bearing was removed from the test installation and dismantled. Subsequent non-destructive testing of the raceways and radii of the slewing bearing (by Magnetic Particle Inspection – MPI and visual assessment) confirmed the ultrasonically assessed spalling with a good correlation. No defect growth (from the initiation defects) was observed.

The detection of spalling had been a possibility in theory, and these tests showed that spalling detection and monitoring was a practical inspection technique for an industrial environment. The authorities that stipulate the dismantled NDT examinations on the slewing bearings accepted the ultrasonic technique (at this stage) as being suitable for this purpose.

Technique Improvements

We recognized that the technique needed improvement, in order to obtain more reproducible amplitude measurements and a more accurate length determination, so that we could monitor defect length changes more accurately.

It was decided to repeat the experiments, using an improved ultrasonic monitoring technique and larger initiation defects (to increase the likelihood of natural crack growth). Accordingly, these larger defects were introduced into the bearing at 180° from the original defects. The slewing bearing was then reassembled with the load placed also at 180° from the originally loaded point. This means that the new loading point coincided with the location of the new initiation defects, and the original defects 11, 12, 21 and 22 would now be very lightly loaded, but defect 3 would be as heavily loaded as during the first series of tests.

An improved ultrasonic recording system was employed for this trial. This consisted of a laboratory ultrasonic instrument package (Krautkramer KS 3000) with an analog output of the amplitude of any signals within the time-gate on the instrument. In addition, a paper chart recorder was employed to record the amplitudes of the signals around the circumference of the slewing bearing. A second channel on the recorder was used to detect signals from an additional proximity detector that sensed the location of the bearing supporting arms as the bearing rotated. This allowed a more accurate location of defects around the circumference during the testing. Figure 3 shows the equipment installed on the factory test machine.

Figure 3 Ultrasonic equipment on factory test-machine

These tests were much more reliable, enabling us to record the entire circumference of each tested area in a much quicker and more accurate way. One restriction of the test technique, however, was the relatively slow response-time of the analog recorder (0.5 seconds full-scale), that tended to "under record" amplitudes (i.e., show them as smaller than they were in real life).

If a constant rotation speed was maintained during the testing, however, this error was reproducible, allowing one recording to be directly compared to any other.

These new tests continued with regular recordings being made from all the tested areas until an additional 21600 rotations had been made. Figure 4 is a typical recording.

At this stage, the testing was again stopped, due to the extent of the new spalling that had occurred. Ultrasonic amplitude recordings when compared, showed very little defect amplitude increases when normalized to a constant sensitivity.

This led us to question if the test system was functioning correctly, as some defect growth from the initiation defects had been expected, but not demonstrated.

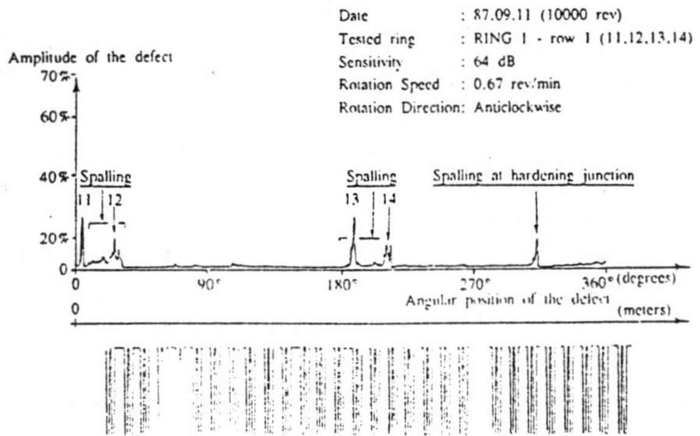

Figure 4 Typical ultrasonic recording

New instrumentation was then purchased (following the order for a slewing bearing with spalling monitoring possibilities) that was both portable and better than the "old" instrument/recording system.

This new equipment consisted of a portable ultrasonic flaw detector, modified to give a 12-bit digitized amplitude output, and a small rugged microcomputer that acts as a system controller and recorder/initial display medium.

The recorded signals can be transferred to a portable ("Lap-Top") Personal Computer (PC), for further signal processing, backups, display and printing. See Figure 5.

The entire system can be operated from batteries, thus being independent of any power supply (with possible mains fluctuations and interference), and the flaw detector/recorder package is suitably rugged, weighs less than 10 kg and can be carried in one hand (an important aspect when climbing ladders in restricted spaces).

Figure 5 Ultrasonic equipment

The time base of the recorded signals can be "zoomed," enabling from 30 samples up to 10000 samples to be displayed across the width of the PC screen. A sample rate of up to 40 samples per second, can now be achieved with this equipment. The normal speed for crane rotation is about 1 to 3 minutes (depending on the installation) for a complete 360 degrees, and this means that for a 1.5 meter diameter bearing, samples at every 0.5 to 1 mm around the circumference can be recorded.

The new equipment was used to make a recording with the factory test installation, and the recorded signals were carefully analyzed.

The two recording techniques produced reasonably comparable results, but the slowness of the analog recorder was clearly observed to have limited the amplitudes of the recorded signals. The new equipment does not have this restriction.

Subsequent non-destructive testing of both the stressed radii and the spalled raceways showed that:

1. The recorded spalling agreed excellently with the spalling observed on the dismantled bearings.

2. No defect growth (on the surface) was observed from the new or the old initiation defects.

Since at this stage, two extensive experiments had been made to get natural cracks occurring in the bearing, and these had failed, despite large initiation defects, it was decided to use an even more drastic method to get natural defects growing in a piece of slewing bearing.

It was decided to take a section of the slewing bearing that contained defect 3, and to install this in a test laboratory, apply a very heavy cyclic loading to this section (to make defect 3 grow if possible) whilst monitoring the integrity of the radius ultrasonically, in an attempt to monitor defect growth.

Experiments on Laboratory Test Specimens
A laboratory set-up was manufactured, that allowed a cyclic load of up to 250 Kilo-Newtons to be applied at a frequency of 7 to 8 Hz.

The length of slewing bearing "nose" that could be expected to fail under 1 million cycles at a 7 Hz cyclic load of 150 K Newtons was calculated to be 75 mm, and consequently, the length of nose studied was reduced to this length by cutting the remaining nose away. Figure 6 shows the set-up used.

During these tests, different ultrasonic recording equipment was employed, namely an XT Personal Computer with an 8 bit A/D converter, sampling the analog amplitude from a Sonatest UFD7 ultrasonic flaw detector. The circumferential translation of the probe assembly (the same probe that was used in both of the initial trials) was performed manually, and is therefore not 100% repeatable nor is the speed constant over the scanned length. The pattern of signals derived from the scans were, however recognizable and fairly repeatable. Graphic representations of the signal amplitudes from the 75 mm long scanned section were displayed on the screen and stored on

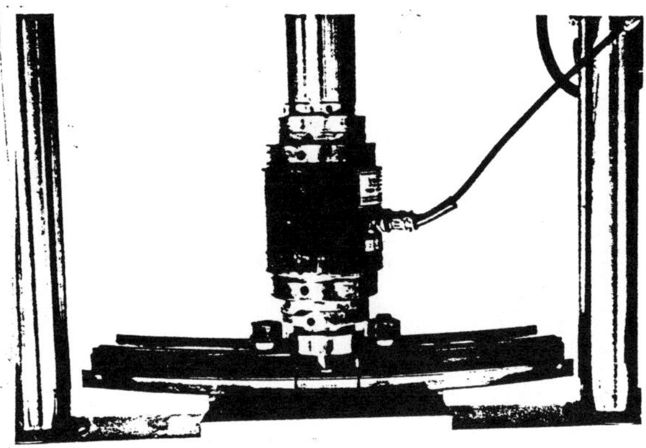

Figure 6 Ultrasonic probe on a portion of slewing bearing "nose"

floppy disk. (Here, about 5 seconds recording time were stored as 1024 consecutive 8 big digitized values, and represent approximately the 75 mm length).

These recordings showed the defect number 3 to be reliably identified throughout the tests.

With a 7 Hz cyclic loading of 70 KN, no apparent defect growth was observed until 500000 cycles. At this point, some apparent growth is seen, but this apparent growth was later shown to be a small variation in the signal amplitudes. From the 500000 to 1000000 cycles, no further "growth" was observed. At this point, it was decided to increase the load to 100 KN, and continue. Here, the next 1000000 cycles failed to produce any noticeable defect growth.

A decision to increase the loading still further to 180 KN and continue with the testing was then taken. Unfortunately, sooner than expected and after only a further 49200 cycles, the bearing nose fractured. This was before the planned ultrasonic test at 50000 cycles. A test revealed a large ultrasonic signal from the radius.

Figures 7a-7f are recordings of the defect signals at this stage, with the number of cycles (from 0 through 1 and 2 million to 2049200).

Figure	Cycles	Load	Remarks
7a	0	0 KN	Start
7b	400000	70 KN	No growth
7c	500000	70 KN	Defect growth suspected
7d	1000000	70 KN	No further growth
7e	2000000	100 KN	No further growth
7f	2049200	180 KN	Fractured section

Tests beyond this point did not reveal any clear indications of defect growth, even though the defect did get larger. This is

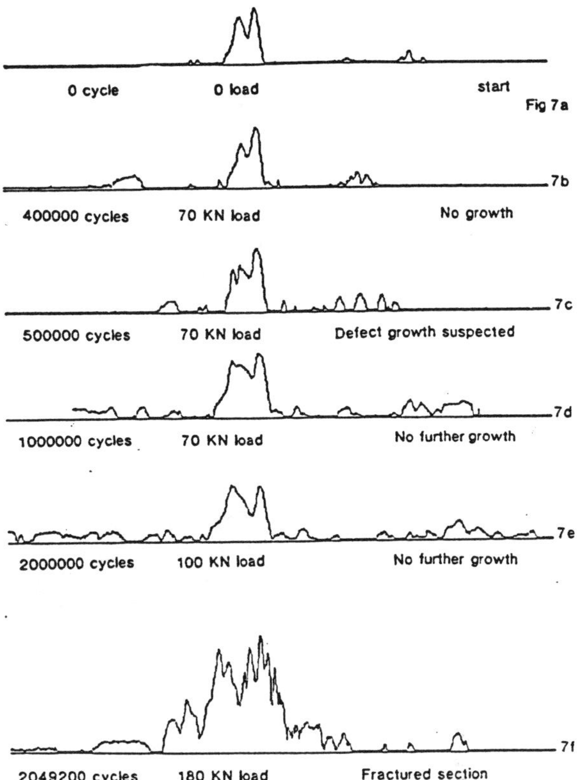

Figure 7a-7f Defect signal recordings at different numbers of cycles

because the defect was extending beyond the main energy lobe of the probe. (Main energy lobe is shown in Figure 2).

It should be noted that the rapid crack growth at 180 KN did not occur from the defect 3, but from a natural crack in the center of the radius, see Figure 8. (Defect 3 was not located exactly at the center of the radius, i.e., the most highly stressed part, but a little to one side, although still on the radius).

RESULTS

These ultrasonic tests showed the following:

1. The crack propagation occurred where predicted.

2. Large growth of the defect is easily recognized, but then only up to a certain size. After this, defect growth occurs outside of the main energy lobe of the probe (see Figure 2). This is not a real drawback to the test technique, since the bearing should already have been labelled as dangerous at a much earlier stage.

3. Small depth growth of defects is difficult to recognize because amplitude evaluation only of the signals is prone to many variables affecting this signal amplitude, and it is difficult to show that the defect has NOT propagated in depth. Only by comparison over several tests can this be confirmed. This is the basis of SKF's monitoring method.

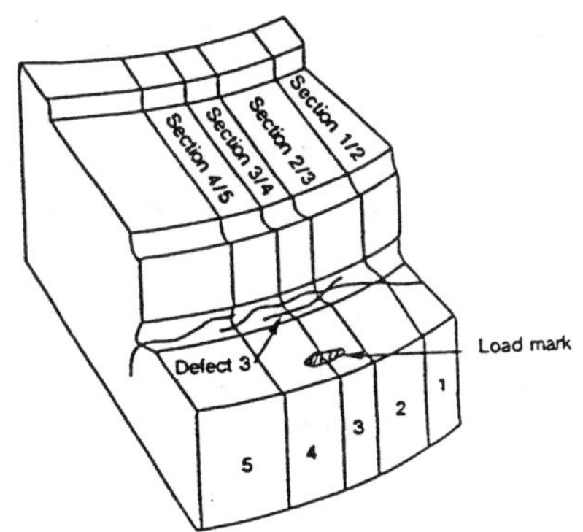

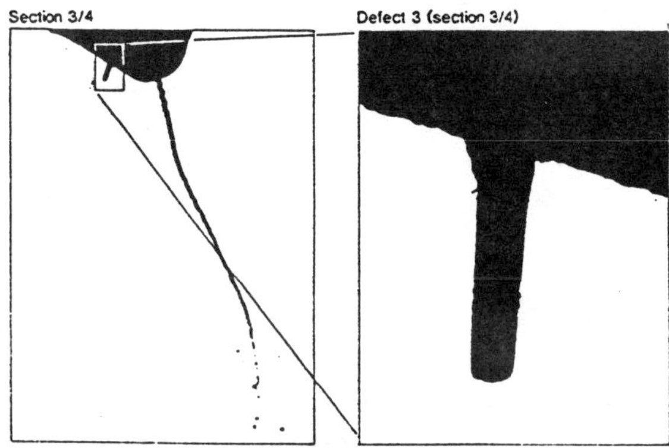

Figure 8 Section location

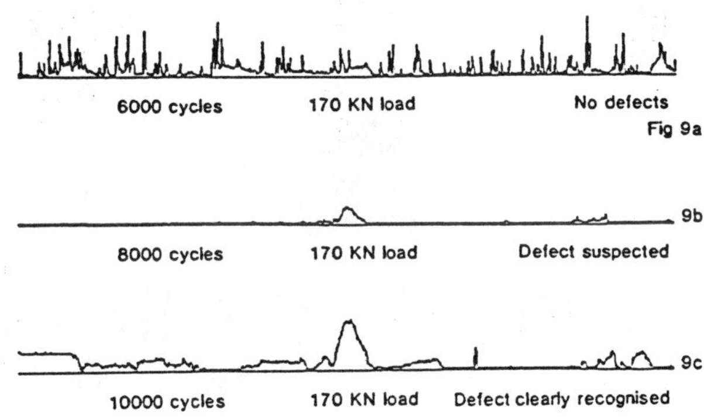

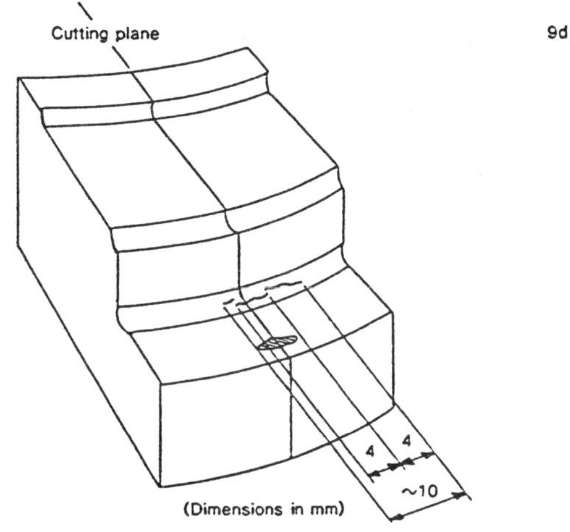

Figure 9 Ultrasonic indications at an increasing number of cycles

4.   We had been able to grow a defect!

A further experiment was now planned and carried out to obtain some ideas about the smallest detectable defect with such a test system. A further (uncracked) 75 mm length of slewing bearing was subjected to a similar test, this time with very frequent tests (every 2000 cycles), but with a cyclic loading of 170 KN. The test was stopped immediately when a defect was recognized from the ultrasonic signal (seen at 8000, clearly recognized at 10000 cycles). Here, the defect can be clearly seen with ultrasonics (Figure 9), but is not detectable with manual magnetic particle inspection (MPI). Even MPI with fluid based fluorescent ink and a x20 magnification microscope to view the radius does not adequately reveal the defect for guaranteed detection.

A metallurgical investigation of this defect revealed it to be three individual cracks, with a total length of approximately 10 mm, and a depth of 0.15 mm.

A further test to show a defect initiating and growing, again on an uncracked piece of slewing bearing, was performed. This time, again 170 KN was used, but the test was continued to destruction of the test piece. Ultrasonic sampling at approximately every 2000 cycles was also used, and in this case, the growth of the defect can be clearly demonstrated.

Figure	Cycles	Load	Remarks
10a	0	0 KN	Start – No defects
10b	12000	170 KN	Crack detectable by US
10c	27500	170 KN	Clearly detectable (US)
10d	45000	170 KN	Clearly detectable MPI
10e	50000	170 KN	Very serious defect

Only at 10d is the defect clearly detectable with MPI, when it is 35 mm long, and shown as a continuous defect on the surface. At 52480 cycles, we had to stop the testing, as the stroke of the testing machine had extended to its limit (due to bending open of the crack).

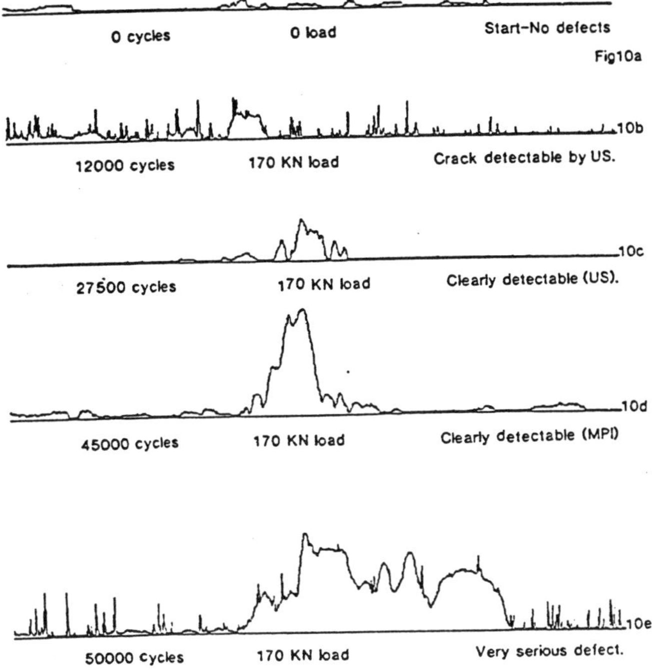

Figure 10 Defect indications at an increasing number of cycles

CONCLUSIONS

A. On the experiments

- In all the laboratory tests, we have shown that the defect always propagated from the predicted position (the center of the stressed radius), and then in the predicted direction. The monitoring technique thus looks at the correct part of the slewing bearing.

- We have shown that the technique is capable of detecting defects in the stressed radii of the slewing bearings at an earlier stage than the technique currently employed (MPI and Visual inspection).

- Generally any growth in length can be measured with some degree of accuracy ($\pm$ 4 to 5 mm).

- Any detected defects can be regularly monitored.

- Defect growth can be more clearly recognized from the pattern of the recorded signals than from the signal amplitudes.

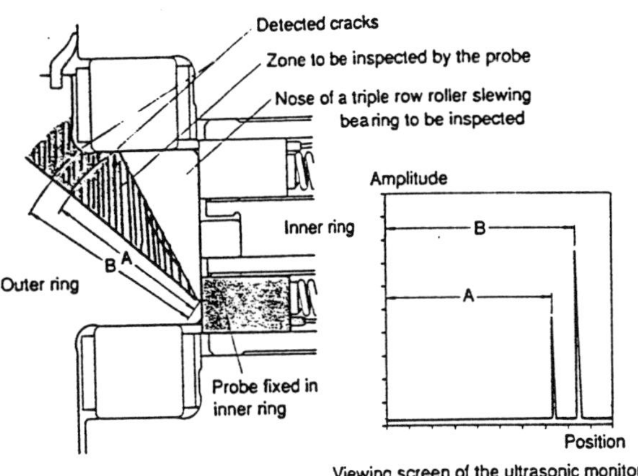

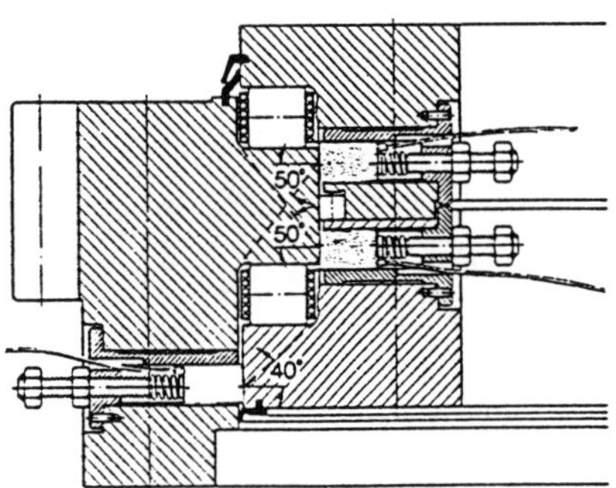

Figure 11 Redesigned slew bearing to facilitate ultrasonic testing

Very small defect propagation cannot be recognized, unless several tests (over a period of time) are compared. Once defects are recognized, more frequent testing MUST then take place. The comparisons will then indicate if there is significant propagation or not. This is only important if a decision to keep the bearing in service is made. The normal life restricting failure of slewing bearings (raceway spalling) can be accurately recorded and followed as the slewing bearing ages.

B. Overall on the technique and application

The new equipment now available can be easily air freighted to exploration and production oil platforms or other cranes, throughout the world.

Under ideal circumstances, the testing system is designed and manufactured into new bearings, however, in many cases this technology can be retrofitted into existing bearings.

The application has been considerably improved since the initial experiments, including some more extensive redesigning of the slewing bearings, to enable the use of more sensitive transverse wave probes throughout the application. A typical application design is shown in Figure 11.

A patent has been granted for this inspection system in the United States of America.

Total applications of this testing system are now 17 new bearings for 6 clients, plus 10 retrofitted systems.1

INSPECTION OF ELEVATOR BRACKETS

Syl A. Viaclovsky
Via NDT, Incorporated
Channelview, Texas

Les Harrington
Les Harrington Consultants
Houston, Texas

In early 1993, Via NDT was called in to evaluate elevator counterweight rail brackets in a high rise building in downtown Houston, Texas. The reason for this was that one of the counterweights in an elevator shaft came out of the rails and struck an elevator causing extensive damage to the elevator car. It happened in the evening while no one was in the elevator. There were no injuries. The elevator was immediately taken out of service until a determination could be made as to the reason for the failure.

Metallurgical and nondestructive testing evaluations were undertaken immediately. A brief summary of the results are discussed in this paper. Due to the proprietary nature of the project, not all of the details could be included. The metallurgical tests were conducted to determine the cause of the failures. In order to determine whether this condition was isolated to this elevator shaft or whether it existed in other elevators, an extensive nondestructive evaluation was conducted. Ultrasonic and visual examinations were performed on all of the elevator shaft counterweight rail brackets.

The brackets supporting the rail system were "L" shaped. The bend radius area of the bracket was where cracking was observed. These brackets hold the rails on which the elevator counterweights move in a vertical direction. This movement can cause cyclic loading of the supports if the alignment and rigidity of the supports is not within the necessary engineering parameters. It was determined that a

failure of the bracket caused the counterweight to jump out of the guide rails which in turn caused the damage to the elevator car.

Metallurgical Evaluation

Tensile tests, notch toughness tests, and chemical analysis tests were performed. The mechanical properties and chemical analysis appeared normal for the steel that was used. However, the notch toughness values using Charpy V-notch specimens, were relatively low, but were not unusual for as-rolled steels. It is necessary, however, to avoid sharp notches and strain intensification points by using generous radii and smooth surfaces. Measures taken in design to distribute the stresses generally have an effect of reducing the initiation of a brittle fracture occurring.

A further detrimental effect of notched material (stress concentrators) is to increase the risk of initiation and propagation of fatigue cracks. This failure appears to have started from a fatigue crack, which began at the indentation formed into the inside curvature of the bracket. The lateral movement of the two rails relative to each other would concentrate that stress at the bend and in particular at the pressed notch made by the forming die. Low loads below static design criteria, can nucleate and propagate fatigue cracks given enough cycles.

Ultrasonic Examination Evaluation

Ultrasonic testing (UT) shear wave examinations were performed on brackets in each elevator shaft. UT shear wave examinations were performed at the location of the radius bends on the elevator rail brackets. These examinations were performed to detect possible cracking in radius bends. A 70° shear wave transducer was used. The UT system sensitivity was set using the reflection from a 0.040-inch deep notch. A representative bracket was obtained and used as the calibration reference standard. Notches were placed along the I.D. and O.D. sides of the bracket (see Figure I). For ultrasonic indication evaluation, a distance-amplitude-curve (DAC) was constructed utilizing the response from the I.D. and O.D. notches in the calibration standard. An 80% response was obtained from the I.D. notch. At this gain setting, the maximum response from the O.D. notch was plotted on the screen. These two points were connected and a DAC drawn on the instrument screen. The results are indicated as a response based on percent of DAC (i.e. response received compared to original DAC).

The UT shear wave examination transmits a sound wave into the material being tested at an angle of 70° from normal. The wave then reflects from the opposite surface or from an indication and returns to the examination detector. If an indication is present, it appears along the time baseline as a vertical deflection. The indication location and length were recorded. Several discontinuity indications were detected by ultrasonic testing. These brackets were replaced.

In addition, fluorescent magnetic particle testing was used to confirm indications found by ultrasonic testing. A substantial number of cracks were confirmed using this method.

Visual Examination Evaluation

Visual examination of the rails in some of the elevators disclosed cracking of the brackets. During visual observation, some of the brackets were also braced to reduce lateral movement of the bracket. The brackets appear to have been designed to allow for vertical movement of the rails relative to the brackets, due to settlement of the structure. In other words, the bracket supports were originally designed to slip if the rails moved up or down. However, no signs of such movement (slipping) were observed. This further contributed to the development of fatigue cracks because of the increased rigidity.

Summary and Conclusions

(1). The metallurgical investigation has shown that the fatigue cracks found were indicative of cyclic lateral loading.

(2). The ultrasonic examination results revealed indications in several brackets. These brackets were removed and replaced.

(3). Visual examination revealed that the brackets and track assembly were not functioning as designed. Engineering modifications were made to alleviate the potential for cyclic fatigue occurring in the bracket bends.

(4). Any sharp notches and strain intensification points should be avoided. The use of generous radii and smooth surfaces was recommended. The replaced brackets did not contain any sharp notches at the bends.

(5). Elevators with a similar design should be examined to determine if cracks exist in the rail support brackets. This will help to ensure that potential failures (counterweight jumps track and strikes the elevator) can be avoided.

Figure 1

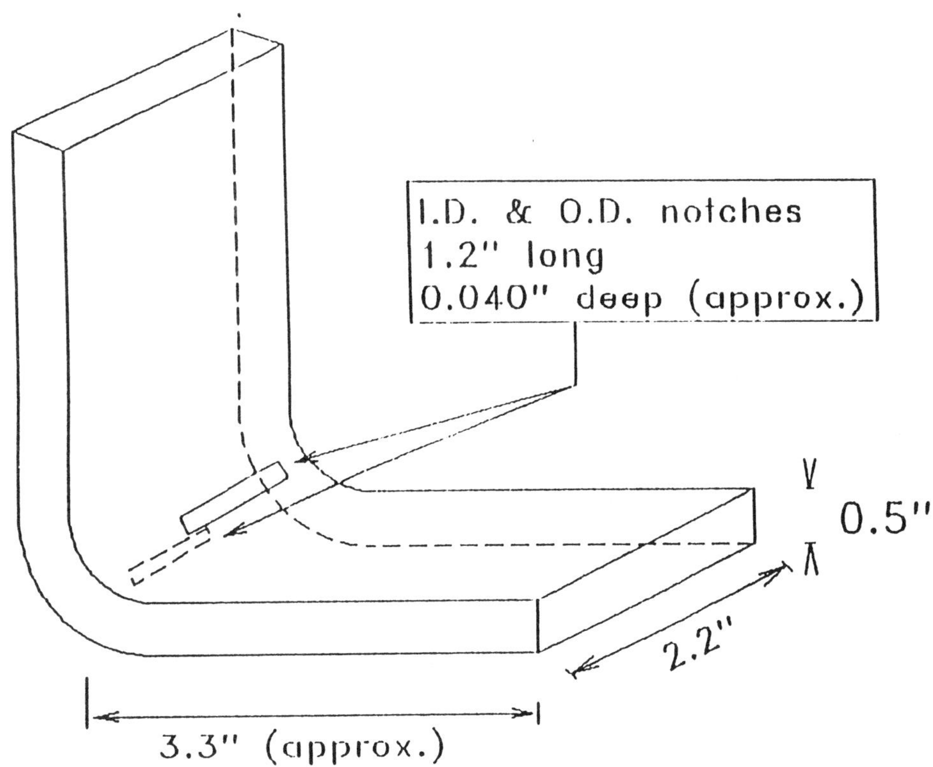

Ultrasonic Examination Calibration Standard

AN ANALYSIS OF UT AMPLITUDE COMPARISON FLAW SIZING AND DISSECTION RESULTS IN STEEL PIPE

Alan V. Bray
Texas Research Institute
Austin, Texas

Roderic K. Stanley
NDE Information Consultants
Houston, Texas

ABSTRACT

An analysis of flaw depth decisions by the ultrasonic amplitude comparison (UTAC) technique is presented. These decisions result from the "prove-up" of a defect discovered during automated inspection. If the maximum depth extent of the flaw is judged larger than a critical depth a decision is made to reject the pipe. Data analyzed were paired UTAC flaw depth decisions and metallurgical flaw depths, the latter taking two forms; data obtained by metallurgical dissection and polishing, or data obtained from grinding through a flaw to uniform metal. The data are analyzed based on the rate of proper/ improper rejection when compared to the "ground truth" metallurgical data. As expected the probability of proper flaw sizing increases with flaw size, leveling out above a given flaw size in a manner similar to classic NDE detection probabilities. This type of analysis appears to lend itself to technique/method evaluation by establishing a flaw sizing characteristic curve for a given method in a given inspection application.

INTRODUCTION

The UTAC inspection method is commonly used to establish whether a longitudinal flaw detected in full length tubular inspection during production meets the depth requirements of the American Petroleum Institute's Specification for Casing and Tubing, SPEC 5CT; Drill Pipe, SPEC 5D; and Line Pipe 5L [1]. It consists of establishing a flaw size criterion by comparing reflected amplitudes from the flaw under inspection and a reference notch of known depth. Typically these results are presented as a percentage of screen height on a scope calibrated such that the reference notch reflected signal is 80% of full screen height. An alternate format is one which uses a previous measurement of the same flaw as a reference, a method often used in verifying flaws prior to grinding [2].

UTAC rejection of pipes for flaw depth is based on the comparison of the reference notch and flaw reflection amplitudes. Typically a reference notch is cut with depth equal to the flaw rejection criteria for the pipe size and grade, usually expressed as a percentage of wall thickness. The statistics of the rejections compared to the actual depth of the flaw are of interest to both producers and consumers of pipes. Rejections which should have been accepted cost the tubular mill by reducing the value of stock, while acceptances which should have been rejected can cost the consuming oil operations in terms of safety and reliability.

FLAW REJECTION DECISIONS

There are two commonly available sources for verifying rejection decisions in UTAC inspections; flaw dissection data from metallurgical examinations and flaw depth data from grinding operations. These absolute flaw depth measurements, combined with the inspection data which resulted in acceptance or rejection, provide a basis for examining UTAC inspection decision accuracies and producer and consumer risks. The sizing inspection, or prove up, results in a rejection or acceptance of the tubular. A subsequent grinding or dissection measurement provides the actual flaw size, which is above or below the rejection level for the pipe size and grade. Figure 1 shows the matrix of possible outcomes. Proper outcomes are those in which an acceptance or rejection is matched with an acceptable or rejectable true flaw depth respectively, AA and RR in the table. Rejections which

<table>
<tr><td></td><td>**Actual Flaw Depth ≤ D**</td><td>**Actual Flaw Depth > D**</td></tr>
<tr><td>**UTAC Acceptance That Depth ≤ D**</td><td><u>AA</u>
Proper Decision</td><td><u>AR</u>
Consumers Risk</td></tr>
<tr><td>**UTAC Acceptance That Depth > D**</td><td><u>RA</u>
Producers Risk</td><td><u>RR</u>
Proper Decision</td></tr>
</table>

FIGURE 1. FLAW SIZING DECISION MATRIX FOR REJECTION DEPTH D

TABLE 1. DECISION OUTCOMES OVER ALL WALL THICKNESSES IN EXAMPLE DATA SET

	All	1/4" Trans-ducer	95% Confi-dence Limits	1/2" Trans-ducer	95% Confi-dence Limits
Proper Depth Decisions (AA&RR)	63%	59%	54-63%	78%	69-84%
Producers Risk (RA)	18%	22%	18-26%	5%	2-10%
Consumers Risk (AR)	19%	19%	15-23%	17%	10-24%
Number of Samples	502	388		114	

should have been accepted, RA outcomes, are at the producers risk, while acceptances which should have been rejected, AR outcomes, are at the consumers risk. In detection theory (e.g.[3]) RA outcomes are called false alarms, i.e., the flaw depth is actually less than the critical depth, but an "alarm" was sounded, indicating it was deeper than the critical depth.

A data set of over 500 sizing measurements, or more properly decisions, and absolute depth measurement pairs was assembled for example purposes. It is difficult to tell whether the data set is typical or not. The data are collected from a variety of sources, including grinding operations, third party inspection services, and mill prove up operations. They include pipe from four mills, at least four operators, and at least three different measurement systems. All data were collected using 2.25 MHz transducers with either 1/4 or 1/2 inch elements. The sample size for the 1/2 inch transducer data was 114 matched pairs, while the 1/4 inch element data included 388 matched pairs.

Operator, equipment type, calibration, technique, and many other factors impact these statistics. The pipe wall sizes available in the data ranged from 0.271 (6.9 mm) to 0.625 (15.9 mm) inches. Over this range of pipe size, and ignoring wall dimension dependence, the decision ratios defined in Figure 1 are presented in Table 1. The upper and lower 95% confidence intervals for proportions are based on the binomial distribution.

As one can guess from the confidence intervals the 1/2 inch element data are significantly different from the 1/4 inch element data at the 95% confidence level. This is not, however, a solid basis for determining that the use of a 1/2 inch element will lead to better decision making, since other factors like operator, sample size, processing equipment, calibration notch length, and wall size dependence are involved.

FLAW SIZE DISTRIBUTIONS

The data of Table 1 ignore the effect of pipe wall on the accuracy of flaw sizing decisions. A system which can accurately size flaws of a given depth would have a different set of acceptance statistics for different pipe walls. As an example a 1.0 inch (25.4 mm) tubular being inspected for 5% flaw depths has an effective rejection depth of 50 mils (1.27 mm). One would expect low inspection error rates if the system were capable of accurately sizing down to 20 mil (0.51 mm) flaw depths. A 0.25 inch (6.35 mm) wall pipe inspected for 5% flaw depths, or 12.5 mils (0.32 mm), would have larger error rates than the 1.0 inch (25.4 mm) tubular if using the same system.

To examine the dependence of these decisions on flaw depth, the cumulative distribution of actual flaw depths and the cumulative distribution of proper decisions can be compared. The actual or absolute flaw depth distribution indicates the percentage of flaws less than a certain depth as a function of depth. The proper decision ratio indicates the number of proper decisions for flaw depths smaller than a certain depth. Figure 2 shows a series of actual and proper decision cumulative distributions for a variety of pipe wall sizes. If all flaw depth decisions were perfect the proper decision curve would overlap the actual flaw depth curves.

The cumulative distributions of Figure 2 provide some insight to UTAC inspection accuracies. Consider the curves in Figure 2a compared to those in 2b. These were measurements made on the same production run using two transducer sizes, 1/4 inch (6.35 mm) and 1/2 inch (12.7 mm), at 2.25 MHz in 0.450 inch (11.4 mm) wall pipe. Notice that the 1/4 inch unit fails to accurately size out to 4-6% of wall thickness, but then tracks the actual flaw size well thereafter. Essentially the two

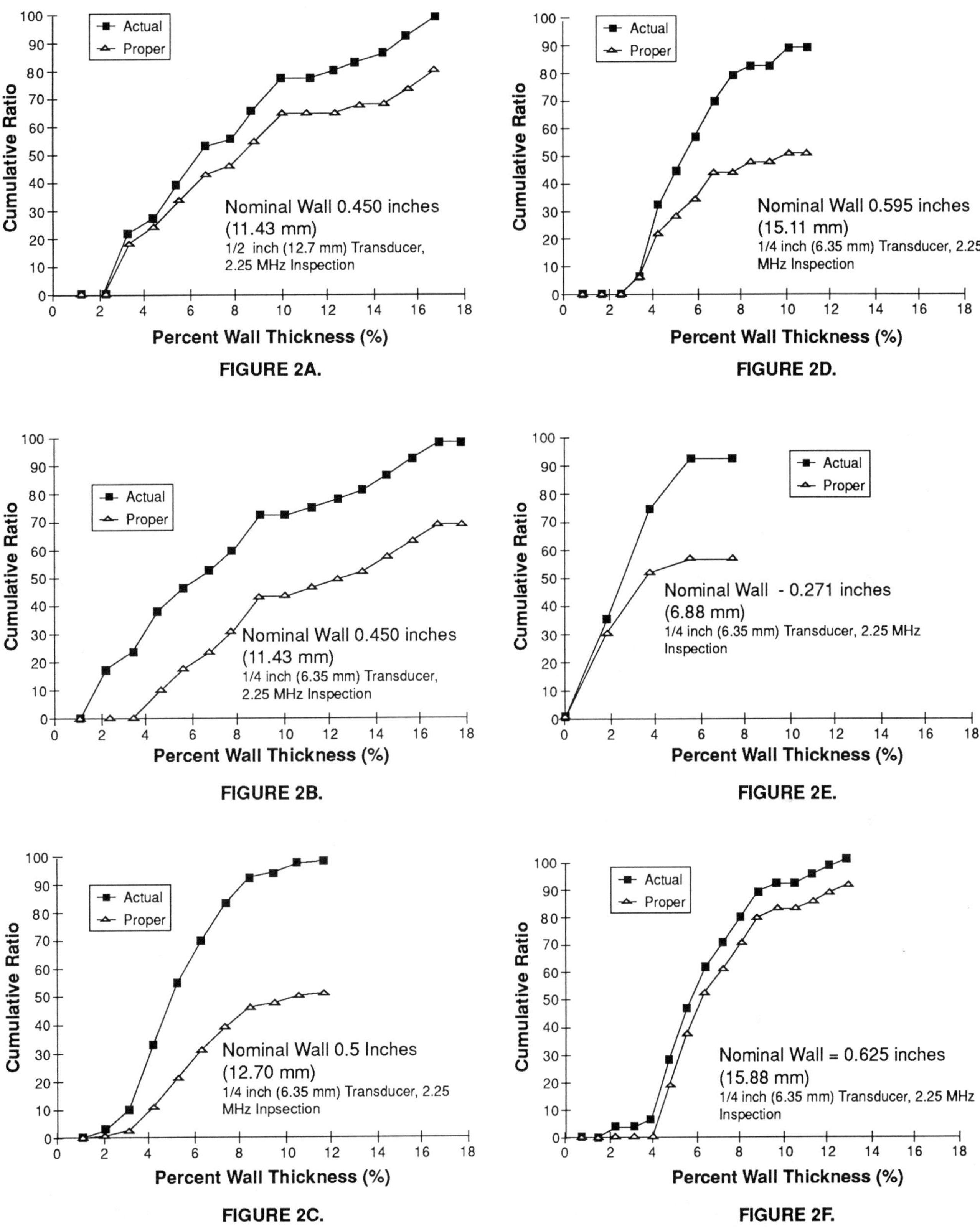

FIGURE 2A.

FIGURE 2B.

FIGURE 2C.

FIGURE 2D.

FIGURE 2E.

FIGURE 2F.

FIGURE 2. ACTUAL FLAW SIZE CUMULATIVE DISTRIBUTIONS AND FLAW SIZE DECISION CUMULATIVE DISTRIBUTIONS FOR A VARIETY OF PIPE WALL SIZES

units produce nearly the same decision accuracy after 5% wall thickness. This is one of the pitfalls of interpreting UTAC data in a cumulative manner, a few early (low flaw size) missed decisions can cause the appearance of the cumulative decision curves to diverge dramatically from the actual flaw population. This same type of problem is evident in the 0.5 inch (12.7 mm) wall tubular of Figure 2c where early misses make the cumulative decision curves diverge, despite nearly perfect tracking after 8% flaw size. Early misses tended to fall into producers risk (as expected, an item rejected when it should have been accepted is more likely to occur in the case of small flaws).

Other cases, like that in Figure 2d for a 0.595 inch (15.1 mm) wall pipe and Figure 2e for a 0.271 inch (6.9 mm) tubular, the smallest wall in the example data, show a poor correspondence in the 4-6% wall thickness range but achieves high proper decisions at larger flaw sizes. Figure 2f shows the prove up data for a 0.625 inch (15.9 mm) which had excellent sizing decision accuracy at a all flaw sizes. In the case of the smallest wall prove up examined, 0.271 inch (6.9 mm), it should be noted that the 5% wall thickness here is 13.6 mils. The largest prove up, the 0.625 inch (15.9 mm) wall data, has a 5% wall thickness of 31.3 mils (0.79 mm).

To get a summary assessment of the decision accuracy of UTAC methods from the data assembled here the impact of factors like early misses and poor large flaw tracking needs to be better described. One approach is to summarize the decision data at a given flaw size. This essentially reduces the problem to one of a series of coin tosses, i.e., at any specified actual flaw size the probability of making a correct decision is 50-50 if the decision is random, like a coin toss. Thus if 50 decisions are made in the 5 to 10 mil (0.13 to 0.25 mm) range and 25 were made correctly, this range of flaw sizes might well have been inspected using a coin toss for each decision. Figure 3 shows the ratio of proper decisions as a function of flaw size in 5 mil increments for the 1/4 inch (6.35 mm) transducer UTAC inspection data. Also shown in the figure are the 95% confidence intervals at each flaw size range. These were computed using the binomial distribution for proportions. The confidence intervals are not necessarily symmetric because of the limits imposed by the nature of the ratio (a zero to 100% range). Figure 4 shows the same data arranged by flaw size as a percentage of wall size.

The dependence of the proper decision ratio on flaw size is clear in Figures 3 and 4, despite some sampling noise in the data. The lower confidence level curves are consistently over 50% after a 30 to 40 mil (0.76 to 1.02 mm) flaw size. Based on the example data and the analysis of Figure 4 it would appear that 1/4 inch (6.35 mm) 2.25 MHz prove ups are capable of accurate sizing, i.e., significantly better than a random choice,

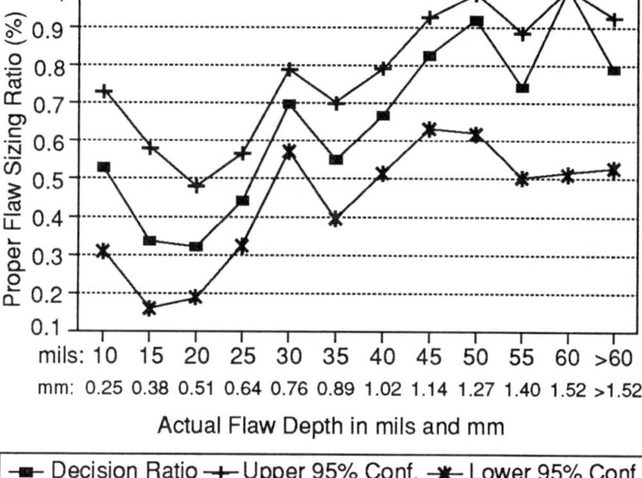

FIGURE 3. UTAC DECISION RATIO VS. FLAW DEPTH, 1/4-INCH (6.35 mm) TRANSDUCER 2.25 MHZ INSPECTION

at approximately 5% of wall in this range of tubular sizes. One would assume that if the process remained unchanged and sampling continued ad infinitum that the upper end of the proper decision versus flaw size curves would be asymptotic to 100%. This level of scrutiny was not possible in the 1/2 inch (12.7 mm) transducer data due to insufficient sample size.

Both figures show an unexpectedly high (>50%) proper sizing decision ratio at the smallest flaw depth. Since all flaws are detected and marked before sizing, these flaws must have had strong enough signatures during inspection to register. In instances where initial inspection is done ultrasonically, these may be those flaws which reflect energy very efficiently for their size. An example would be a flaw which makes an oblique angle to the surface. It is also worth noting that the denominator in the proper sizing ratio is the number of flaws detected (and later dissected) in a given size range. In the lower size ranges this number is small since detections occur at a lower rate than at larger flaw depths. The combination of these factors makes the smallest size range decision ratios in these data questionable.

To examine the producer and consumer risks associated with the data in Figures 3 and 4, Table 2 summarizes the 1/4 inch transducer flaw sizing decisions as a function of flaw size in the example data set. Table 3 summarizes the same data as a function of flaw size expressed as a percentage of nominal pipe wall. Rejection criteria were always 5% of pipe wall. Consumers risk (AR) in Table 3 is highest in the 4 to 10% of wall region, while producers risk (RA) is highest in the 0 to 6% region. In terms of actual flaw depth (Figure 3) consumers risk is highest

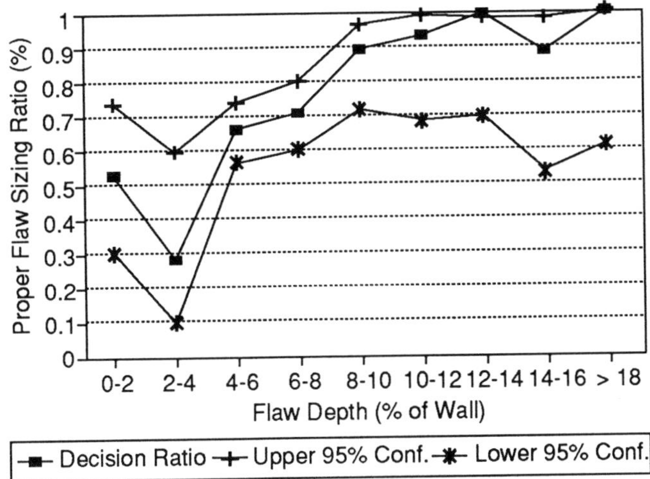

FIGURE 4. UTAC DECISION RATIO VS. FLAW DEPTH AS PERCENTAGE OF WALL, 1/4-INCH (6.35 mm) TRANSDUCER 2.25 MHZ INSPECTION

TABLE 2. DECISION OUTCOMES (%) AS A FUNCTION OF FLAW DEPTH FOR 1/4-INCH (6.35 mm) TRANSDUCER AT 2.25 MHZ IN EXAMPLE DATA

Flaw Depth Range (mils/mm)	AA	AR	RA	RR	n
6-10 / 1.3-0.25	53	0	47	0	17
11-15 / 0.25-0.38	22	6	61	11	18
16-20 / 0.38-0.51	32	5	63	0	41
21-25 / 0.51-0.64	34	0	56	9	64
26-30 / 0.64-0.76	1	26	4	68	72
31-35 / 0.76-0.89	0	46	0	54	46
36-40 / 0.89-1.0	0	33	0	67	48
41-45 / 1.0-1.1	0	18	0	82	28
46-50 / 1.1-1.3	0	8	0	92	12
51-55 / 1.3-1.4	0	26	0	74	19
56-60 / 1.4-1.5	0	0	0	100	9
>60 / >1.5	0	21	0	79	14

TABLE 3. DECISION OUTCOMES (%) AS A FUNCTION OF FLAW DEPTH AS A PERCENTAGE OF PIPE WALL IN EXAMPLE 1/4-INCH (6.35 mm) 2.25 MHZ DATA

Flaw Depth Nominal Wall (%)	AA	AR	RA	RR	n
0 – 2	53	0	47	0	17
2 – 4	28	0	72	0	90
4 – 6	3	23	12	63	137
6 – 8	0	27	3	71	79
8 – 10	0	12	0	89	26
10 – 12	0	7	0	93	14
12 – 14	0	0	0	100	9
14 – 16	0	0	13	88	8
>16	0	0	0	100	8

in the 25 to 45 mil (0.64 to 1.14 mm) range, while producers risk is highest in the zero to 25 mil (0 to 0.64 mm) range.

Table 4 summarizes similar results for the 1/2 inch (12.70 mm) transducer data. These data are difficult to interpret since the sample sizes were low. For small flaw sizes, to about 30 mils (0.76 mm) the 1/2 inch (12.70 mm) transducer data have proper decision outcomes of over 80%, but at higher flaw sizes, from 30 to 60 mils (0.76 to 1.52 mm), the combination of sample size and possibly resolution seems to produce poorer results. Very large flaw sizes (>61 mils (>1.55 mm)) appear reasonable, but again, sample size limits interpretation.

CLOSING REMARKS

The approach described relates flaw depth rejection decisions to actual flaw depth measurements. The analysis, unfortunately, is data intensive. A sample size of 188 paired measurements of 1/2 inch (12.7 mm) transducer 2.25 MHz data were insufficient to draw conclusions about. The 1/4 inch (6.35 mm) transducer example data at the same frequency, which had a sample size of 388, could be examined in detail.

There is a clear flaw size dependence in the 1/4 inch transducer 2.25 MHz UTAC decision data examined. The operant wavelength in these measurements is roughly 85 mils (2.16 mm), and the proper decision ratios get to reasonable levels at about half that value. Clearly the biggest risk for producers is at the lower wall sizes. UTAC decision ratios over 70% are achievable for many pipes inspected at the 5% flaw depth criterion in the example size range. For the smaller wall sizes the producers and consumers risks can be significant.

The UTAC examples used grinding operations and mill dissection data, and included over 500 decision – actual depth pairs. Despite this large data set other factors such as operator competence, type of equipment used, and similar items, make the examples above no more than that – examples. This can be seen in the wide range of proper sizing decisions versus actual size comparisons of Figure 2. Some data were clearly generated by above average quality flaw size estimation methods and or techniques. This is a clear cut case where method and technique need to be standardized to the degree possible to minimize producer and consumer risks.

Flaw Depth Range (mils/mm)	AA	AR	RA	RR	n
11-20 / 0.28-0.51	81	0	19	0	21
21-30 / 0.53-0.76	22	13	9	57	23
31-40 / 0.79-1.02	0	43	0	56	23
41-50 / 1.04-1.27	0	18	0	82	11
51-60 / 1.30-1.52	0	60	0	40	5
>61 / >1.55	0	9	0	91	11

ACKNOWLEDGMENTS

We are indebted to Mill Claim Services, Channelview, Texas for sharing their repair-grinding data with us.

REFERENCES

1. American Petroleum Institute Standards and Specifications 5CT (Casing and Tubing), 5D (Drill Pipe), 5L (Line Pipe), latest editions.

2. Stanley, R.K., "Repair Grinding – An Alternative to Ultrasonic Rejection of Oil Field Tubes," *Materials Evaluation*, Volume 52, Number 19, October 1994.

3. Example: Burdock, S.B., *Acoustic System Analysis*, Prentice Hall, NJ, 1984.

4. Example: Tobias, P.A., and Trindale, D., *Applied Reliability*, Van Nostrand Reinhold, NY, 1986.

AUTHOR INDEX

NDE-Vol. 13
NDE for the Energy Industry – 1995

Book Number: H00930